An Introduction to Seismic Interpretation

R McQuillin M Bacon W Barclay

Published by
Graham & Trotman
London
GULF PUBLISHING COMPANY
Book Publishing Division
Houston, London, Paris, Tokyo

First published in the United Kingdom by
Graham & Trotman Limited
Bond Street House
14 Clifford Street
London W1X 1RD, United Kingdom

This edition published 1979
by Gulf Publishing Company
Houston, Texas, USA

Reprinted in 1980
© 1979 by R. McQuillin, M. Bacon, and W. Barclay

Reprinted 1981
Printed in Great Britain by
King's English Bookprinters Limited
Leeds, Yorkshire.

ISBN 0-87201-797-4

Library of Congress Catalog Card Number 79-7577

CONTENTS

PREFACE

This book was commenced in 1976 when all three principal authors were working in Edinburgh. Chapters one to eight form the main text of the book whereas the final four chapters are case histories which aim to illustrate the application of interpretational techniques to a range of specific exploration problems.

The main text is a joint collaboration between the three authors, with primary responsibility falling to Robert McQuillin. The Moray Firth case study was prepared by M. Bacon, the Rainbow Lake and Hewett chapters by W. Barclay and we are indebted to Esso Australia for approving preparation of the Kingfish study by two of their staff, David McEvoy and Ron Steele. Robert McQuillin undertook the task of editing material for publication.

The authors wish to acknowledge their indebtedness to:
The Director of the Institute of Geological Sciences for his permission to publish a number of I.G.S. photographs and open-file records and publications for some of the illustrations; for approving the participation of McQuillin and Bacon in the publication of this book.
Enserch Canadian Exploration Limited for approval of the participation of William Barclay.
Linda Nisbet for secretarial assistance and Angela McQuillin for her work on preparing the illustrations.

Many companies have given valuable assistance by providing information on equipment, examples of various types of geophysical records and material for illustrations. The addresses of companies referred to in the text are supplied in appendix 2.

R McQuillin
Edinburgh

M Bacon
Legon

W Barclay
Calgary

March 1979

1. SEISMIC WAVES

The object of seismic reflection prospecting is to delineate subsurface geological structures using the ability of some horizons within the earth to reflect sound waves. This chapter sketches briefly the elementary theory of seismic waves, and discusses their behaviour in the real earth.

1.1 Elasticity

Seismic waves are elastic waves propagating through the earth; we therefore begin by briefly considering the basic definition of elasticity. For many materials subjected to small applied forces, Hooke's Law holds good; the deformation produced is proportional to the applied force. Consider a length of rubber being stretched by a weight (figure 1/1b). If the original length of the specimen is L and the extension produced is e, we define the strain for this type of deformation as e/L. For a given weight (ie stretching force) the extension produced will depend on the thickness of the specimen. We therefore define the stress as the force acting per unit area. Then Hooke's Law states:

$$\frac{\text{stress}}{\text{strain}} = \text{constant}$$

so in our case of longitudinal stress this becomes:

$$\frac{\text{longitudinal applied force/cross-sectional area}}{\text{change in length/original length}} = Y$$

where Y is a constant, characteristic of the material of the specimen, called Young's modulus.

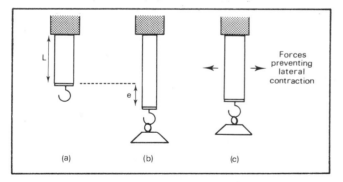

Figure 1/1: Deformation of rubber under applied force.

In the situation of figure 1/1b, when the specimen is stretched not only does its length increase but its width is decreased. We could imagine a situation similar to figure 1/1c when lateral forces are applied to prevent this decrease in width; this is similar to the situation in the solid earth, for when a given section of rock is pulled in a given direction the rocks on either side tend to prevent lateral contraction. The effect of these lateral forces is to decrease the extension under the applied load. In this situation the ratio of longitudinal applied stress to strain is still a constant, called the axial modulus Ψ. We note that the axial modulus will be greater than Young's modulus.

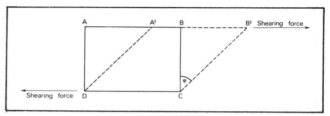

Figure 1/2: Shear deformation.

Another type of deformation is shear, which affects shape but not volume (figure 1/2). The cross-section ABCD is distorted to A'B'CD. Hooke's Law now takes the form:

$$\frac{\text{shearing force/unit area}}{\text{angular deformation } (\varphi)} = \text{constant} = \mu$$

μ is a characteristic of the material and is called the rigidity modulus.

1.2 P and S waves

Elastic waves are of two main types. In longitudinal waves, called P waves by seismologists, the direction of particle motion is parallel to the direction of wave propagation, whereas in transverse waves (S waves) these directions are perpendicular to each other. The P waves are essentially ordinary sound waves.

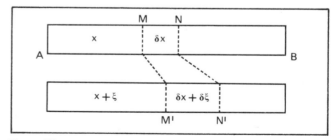

Figure 1/3 Longitudinal waves in a bar.

The example of longitudinal waves in a bar demonstrates how the equation governing these waves can be derived. Figure 1/3 shows a bar AB of cross-sectional area α, of Young's modulus Y, and of density ρ. As a longitudinal wave passes along the rod, each element of the bar vibrates parallel to the rod. The element MN is displaced to a new position $M'N'$. At M' the tension is given by Hooke's Law:

$$T_{M'} = Y \cdot \frac{\text{change in length}}{\text{original length}} = Y \cdot \frac{\delta \xi}{\delta x_{M'}}$$

Thus the force on this element is:

$$(T_{N'} - T_{M'})\, a = Ya\left(\frac{\delta\xi}{\delta x_{N'}} - \frac{\delta\xi}{\delta x_{M'}}\right) = Ya.\delta x.\frac{\partial^2\xi}{\partial x^2}\ ,$$

and since the mass of this element is $a.\delta x.\rho$ and its acceleration is $\dfrac{\partial^2\xi}{\partial t^2}$

$$Ya.\delta x.\frac{\partial^2\xi}{\partial x^2} = a.\delta x.\rho\,\frac{\partial^2\xi}{\partial t^2},$$

ie $\quad \dfrac{\partial^2\xi}{\partial x^2} = \dfrac{1}{c^2}\dfrac{\partial^2\xi}{\partial t^2}$ where $c^2 = \dfrac{Y}{\rho}.$ (1)

Equation (1) is the usual equation of wave motion satisfied by any function of the form $f(x + ct)$, which represents a wave propagating with velocity c.

A similar analysis for the case of a longitudinal wave in the solid earth (a P wave) shows that it propagates with a velocity given by $(\Psi/\rho)^{\frac{1}{2}}$ where Ψ is the axial modulus, because a lateral contraction of the volume element is prevented by the surrounding rock. It can be shown that for S waves, involving displacements perpendicular to the direction of wave propagation, the velocity is given by $(\mu/\rho)^{\frac{1}{2}}$ where μ is the shear modulus.

In seismic reflection prospecting, only P waves are generally encountered. This is because seismic sources generate almost exclusively P waves; in marine work, the source is surrounded by water, in which case the S waves could not propagate even if they were generated. (Fluids have zero resistance to shear and therefore cannot sustain S waves; formally, if $\mu = 0$ the velocity is zero). However, S waves can be generated where P waves strike an interface at inclined incidence, and are therefore of some importance in seismic refraction work (p. 95); in seismic reflection practice, the angles of incidence are normally too small for appreciable conversion to occur.

The amplitude of vibration of the particles transmitting the seismic wave may be extremely small; in the case of reflections from a depth of several kilometres, the amplitude may be only a few angstroms.

1.3 Seismic velocities in rocks and fluid media

We have seen in the last section how to calculate the seismic wave velocity in a single medium. The next step in working towards calculation of seismic velocities in the real earth is to consider a mixture of two components, each of known velocity. Immediately we are in difficulties. At frequencies of interest in reflection seismics, the wavelength of a signal may be several hundred metres, so the seismic wave will not 'see' individual grains of the mixture. A useful starting point is the time-average equation.

$$\frac{1}{c} = \frac{1-\Phi}{c_1} + \frac{\Phi}{c_2},$$

where c is the velocity of a mixture containing a fraction ϕ of material of velocity c_2 in a matrix of velocity c_1. This equation is derived from a model in which the two components of the mixture are physically separated (figure 1/4). In this case the time taken for the seismic waves to transverse a length L of mixture will be:

$$t = \frac{(1-\Phi)L}{c_1} + \frac{\Phi L}{c_2},$$

giving an average velocity c where

$$\frac{1}{c} = \frac{t}{L} = \frac{1-\Phi}{c_1} + \frac{\Phi}{c_2}.$$

It is hard to give much justification for this model, and indeed the time-average equation is of restricted application, as we shall see.

Suppose now that our model earth is still dry, but no longer solid; that is, let it contain empty pore space. A very simple physical model suggests that modest porosity will reduce seismic velocities markedly. In figure 1/5 we divide the rock into columns parallel to the direction of travel of the seismic signal; initially these are in close contact. Porosity is then introduced by moving the columns apart slightly. The average density is reduced, but so is the elasticity because the gaps between the columns contribute nothing to the stiffness of the material. The effects of these changes on the seismic velocity cancel one another out. However, the columns are now able to deform sideways, so we must use Young's modulus rather than the axial modulus in calculating the seismic velocity. Therefore there will be a drastic decrease in velocity. Recollecting that the amplitude of particle displacement in a seismic wave may be only a few angstroms, we see that fractures too small to detect optically could have a marked effect on velocity. The implication is that a very small amount of gas-filled porosity would produce a marked reduction in

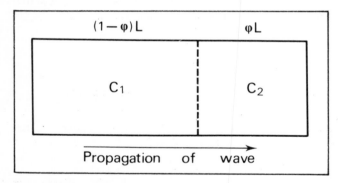

Figure 1/4: Derivation of the time average equation.

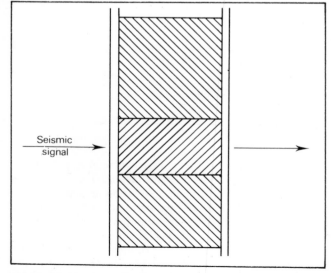

Figure 1/5: Model to illustrate the effect of porosity on velocity in a column of rock.

seismic velocity; this is of great importance for direct hydrocarbon detection (see p.119). Velocities less than that of sound in seawater are sometimes observed in gas-bearing strata.

If the pores are filled with liquid, the situation is more complicated because the liquid tends to resist the lateral deformation of the rock, and the time-average equation may apply. This is true for limestones and for sandstones whose depth of burial exceeds about 2km. At shallow levels, sandstone velocities tend to be unpredictable due to non-consolidation and variable fluid pressure gradients. Thus for a wet sandstone below 2km or a limestone we can write:

$$\frac{1}{c} = \frac{1 - \Phi}{c_1} + \frac{\Phi}{1.5}$$

where c is in km/sec, ϕ is the porosity and c_1 is the matrix velocity, say 5.7km/sec for sandstones and 6.6km/sec for limestones.

Velocities in shales are often well predicted from a relationship between velocity (c) and depth burial (Z) of the type

$$Z = A + B . \ln c$$

where A and B are constants for a given area.

Some typical observed velocities are shown in table 1/1. In general, igneous rocks have higher velocities than sedimentary rocks.

Table 1/1 : Typical ranges of P wave velocities in rocks. (after Grant and West, 1965).

Material	Velocity, km/s
Salt and anhydrite	$4.9 - 6.9$
Granites and metamorphics	$4.0 - 5.8$
Limestone and dolomite	$2.7 - 5.2$
Sandstone and shale	$0.8 - 3.4$

1.4 Densities of rocks and fluid media

In the next section we shall need some knowledge of rock densities. For this case, the theoretical background to the treatment of mixtures and porous material is much more soundly based than for velocities. As in the case of the derivation of the time-average equation, the model requires the separation of the constituents into two partitions of the volume; the validity of this procedure is not in doubt for densities. We then derive that the average density ρ is given by

$$\rho = (1 - \Phi)\rho_1 + \Phi\rho_2$$

where a fraction ϕ has the density ρ_2 and the matrix has density ρ_1. This can obviously be extended to any number of constituents.

For shales, a very significant increase of density with depth occurs; this compaction is largely due to rearrangement of clay particles and is not recovered if the pressure is subsequently reduced by removal of the overburden. As compaction becomes complete, the density tends to about 2.3 g/cc.

Table 1/2 : Typical ranges of Rock Densities.

Material	Density, gm/cc
Igneous rocks	$2.5 - 2.9$
Limestone	$2.3 - 2.8$
Shale	$2.0 - 2.7$
Sandstone	$2.1 - 2.6$
Salt	$1.9 - 2.1$

Some typical observed densities are shown in table 1/2.

1.5 Reflection of seismic waves

If a plane seismic wave strikes a plane interface between two different materials, it will be partially reflected and partially transmitted. For the case of normal incidence we can calculate the reflection coefficient (ie the ratio of the amplitudes of the reflected and incident waves) as follows. If we consider a sinusoidal wave of frequency v incident at the interface (figure 1/6), it will be described by the equation:

$$\xi = \xi_I \exp 2\pi i v(t - x/c_1)$$

where ξ is the displacement at time t, the distance x being measured perpendicular to the interface.

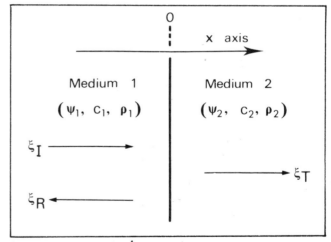

Figure 1/6: Calculation of reflection coefficient at normal incidence. ξ_I, ξ_T and ξ_R are the displacements of incident, transmitted and reflected waves. ψ_1, c_1 and ρ_1 are the axial modulus, velocity and density of medium 1; ψ_2, c_2, ρ_2 those of medium 2.

The reflected and transmitted waves will be:

$$\xi = \xi_R \exp 2\pi i v(t + x/c_1)$$

and

$$\xi = \xi_T \exp 2\pi i v(t - x/c_2) \text{ respectively.}$$

At the interface both displacement and pressure must be continuous. Therefore, if the interface is at $x = 0$, the displacement condition gives

3

$$\xi_I + \xi_R = \xi_T \qquad (2)$$

and, recollecting that the pressure is given by $\Psi . \frac{\partial \xi}{\partial x}$ we see that the pressure condition gives

$$-\frac{\Psi_1.\xi_I}{c_1} + \frac{\Psi_1.\xi_R}{c_1} = -\frac{\Psi_2.\xi_T}{c_2}. \qquad (3)$$

Combining (2) and (3) gives:

$$\frac{\xi_R}{\xi_I} = \frac{\Psi_1/c_1 - \Psi_2/c_2}{\Psi_2/c_2 + \Psi_1/c_1}.$$

But $\qquad c = \sqrt{\Psi/\rho}$, so $\Psi/c = \rho c$.

Therefore $\qquad \dfrac{\xi_R}{\xi_I} = \dfrac{\rho_1 c_1 - \rho_2 c_2}{\rho_1 c_1 + \rho_2 c_2}.$ $\qquad$ **(4)**

This is the required reflection coefficient. It is useful to define the acoustic impedance, r, by

$$r = \rho c .$$

A reflected wave will be observed whenever there is a change of acoustic impedance at an interface. The reflection coefficient for pressure amplitudes will be given by:

$$\frac{P_R}{P_I} = \frac{r_2 - r_1}{r_2 + r_1}.$$

where P_I is the pressure amplitude of the incident wave and P_R that of the reflected wave.

There will thus be a phase reversal (negative coefficient) if the incident ray is in the higher impedance material.

Some typical reflection coefficients at normal incidence are as follows:

Sea floor 1/3
Sea surface -1 (wave incident from below).
Normal strong reflector 1/5

At inclined incidence, derivation of a general formula is difficult because partial conversion into S waves can occur. The reflection coefficient increases as the angle of incidence increases, becoming large in the vicinity of the critical angle, arcsin c_1/c_2. In seismic reflection practice, data is obtained only at near-normal incidence. Data at inclined incidence would be useful where there is no difference of acoustic impedance across an interface because c and ρ change in opposite directions. Equation (4) shows that there will be no reflection at normal incidence, but a reflection would be expected at inclined incidence. This lack of acoustic impedance contrast, despite velocity changes, is sometimes observed at salt horizons.

That part of the energy not reflected at the interface is transmitted into the second medium, undergoing refraction at the interface; the angle of refraction is related to the angle of incidence by Snell's Law, sin r/sin i = c_2/c_1 where c_2 is the velocity on the refracted ray side and c_1 that on the incident ray side, i and r being the angles of incidence and refraction.

1.6 Absorption

Energy is lost by the seismic wave during its transit through the earth. Generally, there is a constant fractional energy loss per cycle of the seismic wave. Some values are as follows:

	Energy loss, dB/wavelength
Weathered rock, gas sands	3
Normal rock	0.5
Lowest observed	0.1

Since there is a constant fractional loss per wavelength, higher frequencies are attenuated more than lower frequencies for a particular path length. The way in which the earth behaves as a low-pass filter can be illustrated by an example. Suppose we have a reflector at a depth of 2 seconds (two-way travel time). In normal rock, the energy of the reflected signal is affected by absorption as shown in table 1/3. There is clearly a very significant loss of high-frequency energy. Even if our seismic input signal had a sharp spike wave form, so that all frequencies were equally represented, it would become a low frequency (say several tens of Hz) signal after traversing a path through the earth typical of reflection prospecting.

Table 1/3 : Acoustic loss in 2s path length due to 0.5dB absorption per wavelength.		
Frequency Hz	*No of wavelengths in path*	*Loss dB*
10	20	10
20	40	20
40	80	40
80	160	80

1.7 Diffraction and interference

Consider a point reflector, which reflects seismic energy back along its incoming path whatever the angle of incidence (figure 1/7). The travel time will be:

$$2(X^2 + Z_0^2)^{1/2}/c.$$

Therefore, if the reflection time is incorrectly assumed to come from a point vertically below the source, the point reflector will produce an apparent event at D, where AD = AB. The locus of such points is given by:

$$z = \frac{2(x^2 + z_0^2)^{\frac{1}{2}}}{c}$$

ie $\qquad c^2 z^2 = 4(x^2 + z_0^2)$

which is a hyperbola whose apex is situated at the point reflector. An actual example is shown in figure 1/8.

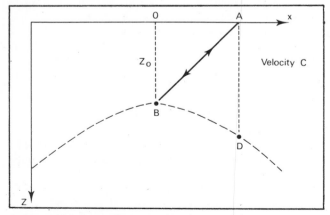

Figure 1/7: Apparent seismic event (dashed line) associated with reflection from point B.

Detailed theory of the seismic wave shows that where a reflector is suddenly terminated (eg by a fault), the end of the reflector behaves like a point source (figure 1/9). In practice, only the section BD of the hyperbola is observed. Such hyperbolic events at faults are common on seismic records (figure 1/10). This is an example of diffraction; the wave interacts with obstacles in ways different from those predicted by simple ray theory.

Another effect stemming from the wave nature of the seismic disturbance is interference. When reflections are obtained from two closely-spaced horizons, the reflected pulses overlap and it may not be possible to separate the two horizons on the seismic record. Since the wavelength of a typical seismic signal will be several hundred metres, this situation is very common and is often found in structures of hydrocarbon significance, such as wedge-outs and fluid contacts. Often such composite events can be identified by careful study of the wave-form.

1.8 Elementary propagation theory

An alternative approach to the description of how seismic waves are propagated through a medium is based on the concepts of geometrical optics, and although this topic is covered in many standard physics text books, it is useful to consider here how these principles help in the understanding of and interpretation of seismic travel times. The propagation of seismic waves through a medium causes a displacement of the individual particles within it. If we consider sinusoidal oscillations, as described in the one-dimensional case by the relation $\xi = \sin(x + ct)$, where ξ is the particle displacement and c is the seismic velocity we see that particles at different distances from the source (differing x values) are at different stages in their oscillations at any time t; that is, they have different phases. One of the ways in which the propagation of a wave may be followed is by joining together the points where the particles have a particular phase to form a wavefront. For example, if we have a point source in a uniform medium the wavefronts will be spheres centred on the source; or, in the one-dimensional case, the wavefronts will be planes perpendicular to the x axis. A wavefront of given phase will advance with velocity c; thus in the one-dimensional case a wavefront situated at $x = x_1$ at time $t = t_1$ will at time $t = t_2$ be situated at x_2 such that a point on the wavefront will have a constant phase, i.e.

$$x_1 + ct_1 = x_2 + ct_2$$
$$x_2 - x_1 = c(t_1 - t_2).$$

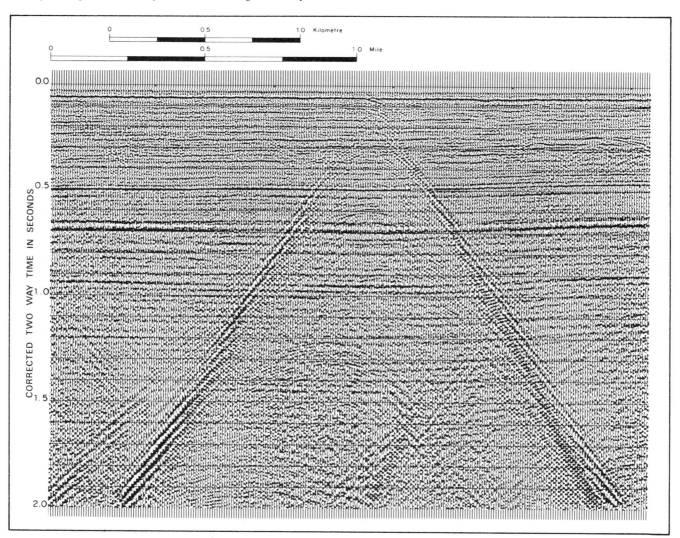

Figure 1/8: Hyperbolic event from a single-point reflector.
(*Courtesy: IGS, S&A record*).

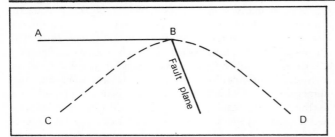

Figure 1/9: Theoretical diffraction pattern from a fault.

The wavefront is propagating in the negative x direction at a velocity c.

The mechanism by which a disturbance is propagated is that each vibrating particle exerts a force on its neighbour, causing it to vibrate. This suggests a way in which, given its initial position, the successive positions of a wavefront can be constructed: we suppose that each vibrating particle on the wavefront behaves as a point source of secondary wavelets. This is called *Huygen's Principle.* The subsequent disturbance is found by combining all these secondary wavelets; in order to explain the observed propagation it is

necessary to suppose that the secondary wavelets produce an appreciable effect in the forward direction only.

A different method of depicting the propagation of the seismic disturbance is to follow the paths along which energy travels. These paths are called rays. In isotropic media the rays are perpendicular to the wavefronts in general. It is often easier to think of wave propagation in terms of rays; see for example the complex reflection pattern produced by a sharply concave reflector as illustrated in figure 3/26. The laws governing ray behaviour are simple, but ray theory is far from adequate to explain all observed propagation effects.

As described in section 1.5, when a plane seismic wave strikes a plane interface between two different materials, it will be partly reflected and partly transmitted. The geometry of this situation is illustrated in figure 1/11. As each element of the wavefront AB reaches the surface A_1B_2, it is a source of secondary wavelets.

In the reflection case, A_1 acts as a source when B is at B_1, so that by the time B gets to B_2 the secondary wavelet from A_1 has expanded to a radius A_1A_2 which also equals B_1B_2. By treating all parts of the wavefront in this way one can construct A_2B_2 as the reflected wavefront. The ray paths have the simple geometry that they are in the same

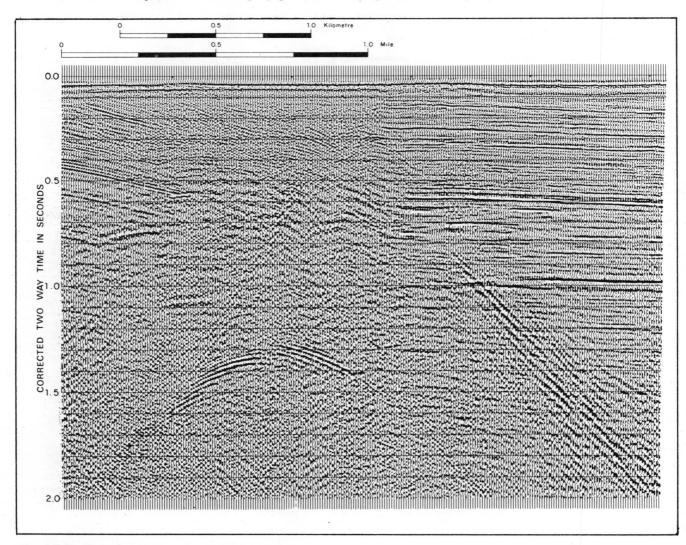

Figure 1/10: Hyperbolic diffraction patterns caused by faulting. (*Courtesy: IGS, S&A record*).

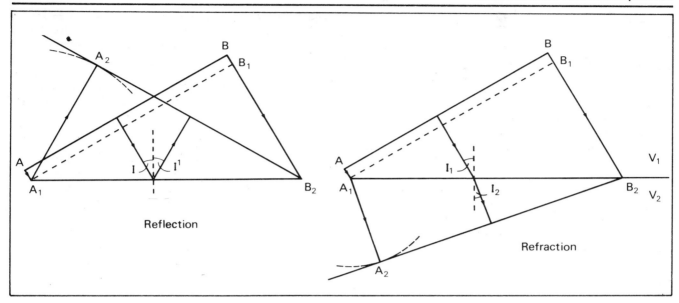

Figure 1/11: Reflection and refraction of a plane wave AB at plane interface A_1B_2.

plane as, and are equally inclined to the normal to the reflector at the point of incidence, i.e. $I = I'$

In the case of the transmitted wave, we see that while the disturbance travels from B_1 to B_2, the secondary wavelet from A_1 expands to $A_1A_2 = B_1B_2,(V_2/V_1)$; in the illustration $V_1 > V_2$ thus the distance A_1A_2 is less than B_1B_2. From the above it can be deduced that the transmitted wavefront is normal to a ray whose direction is given by:

$$\frac{\sin I_1}{\sin I_2} = \frac{V_1}{V_2}$$

The wavefront is said to have been refracted at the interface, and as before, the two wavefront normals and the normal to the refracting surface lie in the same plane. This description of the refracted wave is generally known as *Snell's Law* as noted in section 1.5.

In more complicated cases the behaviour of rays can be predicted from *Fermat's Principle* which states that the ray path along which a disturbance travels from one point to another will be that taking the least time.

1.9 Summary

1. Seismic prospecting utilises reflection of sound waves to delineate horizons within the earth.
2. A reflection occurs wherever there is a change in **acoustic impedance** (defined as the product of seismic velocity and density). The greater the impedance contrast, the stronger the reflection. Such impedance changes normally correspond to lithological variations.
3. Absorption causes the earth to act as a low-pass filter, making deep reflections rich in low frequencies compared to the input signal.
4. Hyperbolic diffraction patterns on the seismic section are generated by point reflectors and by sharp geological discontinuities such as faults and steeply inclined flanks of monoclines, synclines and anticlines.

References and suggested reading

B.S. Evenden, D.R. Stone and N.A. Anstey, *Seismic prospecting instruments.* (Gebruder Borntraeger, Berlin)
Vol. 1: *Signal characteristics and instrument specifications* by N.A. Anstey (1970).
Vol. 2: *Instrument performance and testing* by B.S. Evenden and D.R. Stone (1971).
These volumes give a full account of seismic instrumentation design and performance.

A.A. Fitch, *Seismic reflection interpretation* (Gebruder Borntraeger, Berlin, 1976). A book on seismic interpretation written at three levels: an introductory level, a professional level and a research level.

F.S. Grant and G.F. West. *Interpretation theory in applied geophysics* (McGraw-Hill Book Company, New York, 1965). A mathematical text on interpretation methods used in applied geophysics.

Seismograph Service Corporation, *The Robinson-Treitel Reader* (SSC, Tulsa, Oklahoma, 1973) A collection of papers on digital data processing compiled by SSC as a service to industry.

R.E. Sheriff, *Encyclopaedic dictionary of exploration geophysics* (Society of Exploration Geophysicists, USA, 1973). A well illustrated encyclopaedia covering the whole field of exploration geophysics.

2. DATA ACQUISITION

In this chapter we consider the application of seismic theory, as described in chapter 1, to the design of equipment for generating seismic pulses, and detecting and recording the earth's response to the passage of seismic waves through it. The aim here is not to treat the subject of data acquisition at the level which would be required in a text-book written for geophysicists principally concerned with field techniques, but to present enough background information on this subject to satisfy the needs of the seismic interpreter who may have little opportunity to participate in field surveys, or at best gain experience of only a limited range of the methods currently in use. Exploration seismology is a remote-sensing technique in which the aim is to record as detailed a picture as possible of subsurface geology. The product of a seismic investigation is a geological model which can be described as the sum of a finite series of layers of varying thickness, physical properties (density and seismic velocity) and structural attitude. Interpretation of this model is in terms of geological structure, lithological variation, stratigraphy and, in oil exploration, hydrocarbon prospectivity.

Seismic data is acquired using a system consisting of three main components: an input source, an array of detectors and a recording instrument. The input source is designed to generate a pulse of sound which meets, as near as possible, certain predefined requirements of total energy, duration, frequency content, maximum amplitude and phase. Reflected and refracted seismic pulses (the output from the earth) are detected by an array of geophones or a hydrophone array, then recorded by a recording instrument, and in both cases these output signals will be modified by the response characteristics of that part of the system. Each seismic record is thus a time record of the output signals which are generated at interfaces in a series of stratigraphic layers because of the changes in acoustic impedance which occur at such boundaries, modified firstly by transmission decay and noise interference in the earth and then by detector and recorder response characteristics. This can be summarised as follows:

Recorded signal = Source pulse * [Reflectivity * (Earth filter + Noise)] * Detector response *Recording instrument response,

where * represents convolution (see p.188)

Assuming that we know the signal characteristics of the seismic pulse and the response characteristics of detector and recording instrument, then we can separate that part of the function contained in square brackets, and this is the earth's impulse response. The earth's reflectivity is what we wish to measure. The earth's filter is a variable function of absorption and attentuation which can be compensated for in data processing. Noise cannot be so adequately treated by data processing and, as far as possible, must be measured and compensated for during data acquisition. This is mainly achieved by layout design, on land by proper design of geophone spreads and arrays and at sea by use of well designed hydrophone arrays. Display of the recorded signal will not, in most situations, give an easily interpretable picture of geological structure. This record needs further processing to achieve such clarification, and these processing techniques are the subject of the following chapter.

2.1 Layout design

In designing layout systems, the emphasis is placed on eliminating unwanted signals or 'noise' of both the random and coherent variety. Use of multiple sources, multiple detectors per trace, and the summing of common reflection point traces (see figures 2/1 and 3/6) brings about a distinct improvement in signal to noise ratio in the case of random noise. For spatially random noise, the improvement is proportional to $\sqrt{n}$ where n is the number of detecting elements in the acquisition system, the signals from which are added together to provide the final record. For example, the summing of eight separate seismic signals (eight shots at same shot-point location), detected by geophone spreads of twenty geophones per trace then subject, during processing, to twenty-four fold stacking, will provide an improvement ratio of $\sqrt{8 \times 20 \times 24} = 62$ or 36dB. This may be compared with a single shot record, single-fold processed with, as before 20 geophones per spread, in which case the improvement ratio is $\sqrt{20} = 4.5$ or 13dB. The former acquisition method shows a relative improvement of 23dB over the latter in signal to noise ratio enhancement.

Figure 2/1 shows a typical marine multi-channel acquisition system and illustrates how data are acquired in a way which allows stacking during processing. Although this illustration shows only a marine system, acquisition of land data is based on identical principles. At sea, a survey ship tows a hydrophone streamer made up of a number of sections, numbered one to forty-eight in the figure. Modern streamers are fitted with 24, 48 or 96 such sections and each section consists of a group of hydrophones which are pressure sensitive sound detectors (see p.26) Signals received by the hydrophones in each section are summed so that each section is considered to be an independent single detector. In figure 2/1 the reflections from a single horizon are schematically portrayed as received in the first eight sections of the 48-section streamer. Let us assume that the distance between sections is 50m and that the ship is travelling at 8km/h (approximately 4kts). If the first shot S_1 occurs at time t_1, a reflection from depth point no.1 is received in section no.1 of the streamer and thence

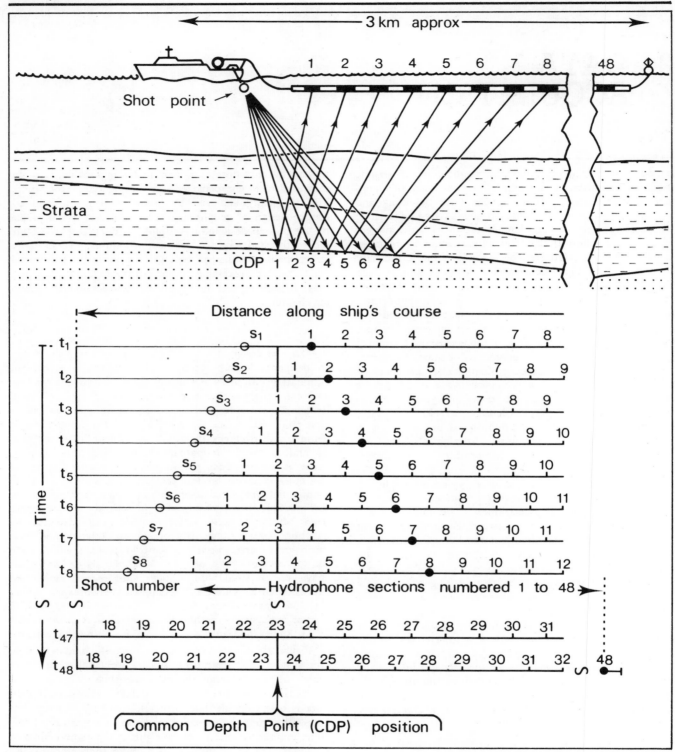

Figure 2/1: Schematic diagram showing the use of multi-channel hydrophone streamers to acquire data which can be common depth point (CDP) stacked.

recorded in channel no.1 of the seismic recording system onboard ship. The common depth point (CDP) position is located mid-way between the locations of S_1 and section no.1; in the lower half of the diagram at t_1 the section no.1 location highlighted as a large dot. The next shot S_2 is timed so that the ship has progressed to a position such that the location of the midway point between source and section no.2 of the streamer is the same CDP location as for S_1. This occurs at t_2, see lower half of diagram, and it can be seen that the distance between S_1 and S_2 is half the distance between sections, that is 25m. The interval between shots $(t_2 - t_1)$ should be set therefore at 11.25s.

At t_2, the signals recorded on channel no.2 of the recording system are therefore those associated with CDP no.1. Shots S_3 to S_{48} follow at the same interval and successive records are obtained on ship from CDP 1, until data from shot S_{48} is recorded on channel no.48 of the recording system. It should be noted that as the ship progresses along course, the seismic signal reflected from CDP 1 will have travelled an ever increasing distance between shot-point and receiving streamer section. The change in geometry is corrected for during processing, and it is possible to add together (stack) all 48 records pertaining to CDP 1. Obviously the same is true for the locations CDP 2, 3 etc. CDP stacking is valuable not only as a means of increasing signal to noise ratio but also, during processing (see chapter 3), of allowing differentation between primary reflections from geological structure, and multiple reflections in sea and rock layers. Multiple reflections can then be suppressed to improve the quality of the final seismic section display. In land surveys, shot-point locations are surveyed at fixed intervals and groups of geophones are pegged into the ground with a group interval which is equivalent to the section interval of the marine streamer. Obviously, surveys on land cannot be conducted with the speed of a marine survey, and timing of shots is irrelevant to a static layout, nevertheless, the geometrical principles are identical.

Coherent noise, of the types illustrated in figure 2/2, can be reduced in two ways depending on whether it is a direct near horizontal wave originating near ground level, in the sea or near seabed, or is a reflected near vertical travelling wave. In general, lowest velocity direct waves arrive latest, a factor which can be utilised in design criteria for detector arrays. With reflected noise waves, lowest velocity waves arrive earliest and this can be utilised in the design of optimum trace spacing for attenuation stacking.

Multiple, or secondary, reflections (as opposed to the primary reflections on which data are being sought) can be attenuated or even effectively eliminated by common reflection point (or common depth point, CDP) stacking. The principle is to design the trace spacing such that the secondary reflections have the appropriate residual normal moveout to be stacked out of phase and consequently much reduced in amplitude, while the primary reflections are stacked in phase by application of the correct normal moveout velocity (see chapter 3). The formulae to be utilised for the simplest cases (see figure 2/3) are as follows:

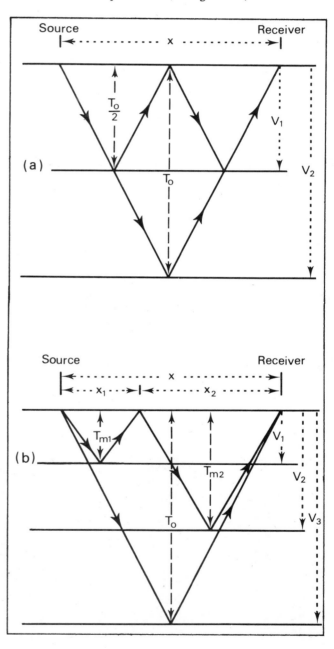

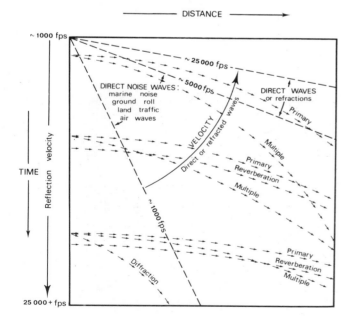

Figure 2/2: Coherent seismic noise types. For primary and multiple reflections the average velocities increase and the normal moveouts decrease with time (after S.D. Brasel in unpublished report, *Design of seismic field techniques*. Atlantic Richfield, 1971)'.

Figure 2/3: Ray path multiple reflection geometry (not adjusted for refraction according to Snell's Law).
a) First order or simple multiple.
b) Second order or peg leg multiple, that is, $T_0 = T_{m_1} + T_{m_2}$. Such multiples are common in marine survey data due to repeated reflections between seabed and sea surface.

i. The residual moveout first order (symmetrical) surface multiple equation (see figure 2/3a for definition of symbols),

$$\triangle t = \triangle T_{multiple} - \triangle T_{primary}$$

$$= 2\left(\sqrt{\frac{T_0^2}{4} + \frac{x^2}{4V_1^2}} - \frac{T_0}{2}\right) - \left(\sqrt{T_0^2 + \frac{x^2}{V_2^2}} - T_0\right)$$

$$= \sqrt{T_0^2 + \frac{x^2}{V_1^2}} - \sqrt{T_0^2 + \frac{x^2}{V_2^2}}$$

ii. The residual moveout second order (asymmetrical) or peg-leg multiple equation (see figure 2/3b for definition of symbols,)

$$\triangle t = \triangle T_{M1} + \triangle T_{M2} - \triangle T_{primary}$$

$$= \sqrt{T_{M1}^2 + \frac{x_1^2}{V_2^2}} - T_{M1} + \sqrt{T_{M2}^2 + \frac{x_2^2}{V_2^2}} - T_{M2}$$

$$- \sqrt{T_0^2 + \frac{x^2}{V_3^2}} + T_0$$

$$= \sqrt{T_{M1}^2 + \frac{x_1^2}{V_1^2}} + \sqrt{(T_0 - T_{M1})^2 + \left(\frac{x - x_1}{V_2^2}\right)^2}$$

$$- \sqrt{T_0^2 + \frac{x^2}{V_3^2}}$$

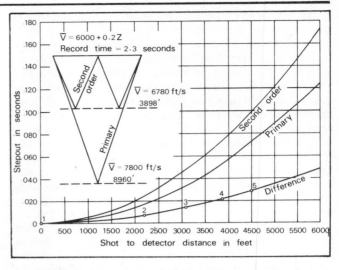

Figure 2/4: Normal moveout differentials between time-co-incident primary and multiple reflections plotted against length of spread. Detector spacing across the spread could be designed to stack the second order signals (by current convention; the first order multiples) out of phase after applying normal moveout corrections to the primary signals..

For optimum attenuation, the trace distances are chosen such that the $\triangle t$ differences are approximately equal to the multiple period divided by (fold of stack-1). Figure 2/4 is adapted from the original paper on stacking by Mayne* demonstrating the principle. In practice, unless expanded spreads are shot (see figure 2/5) knowledge of the type of multiple to be suppressed, and of their velocities, is usually

wanting, and non-uniform trace spacing is impracticable, so the design adopted is invariably a compromise offering limited attenuation. However, it is important to be aware of the factors which control attenuation of multiples, in particular if an area is to be resurveyed with the aim of acquiring improved data in a situation where multiples are known to pose interpretation problems. Figure 2/6 is an illustration of a single-fold as against multi-fold comparison; the distinct multiple suppression and improvement in the signal to noise ratio of the primary reflections is apparent.

* W. Harry Mayne, 'Common reflection point horizontal data stacking techniques'. *Geophysics*, vol. 27, (1962) pp.927–938.

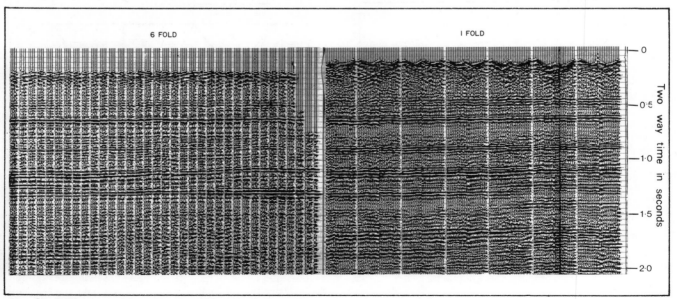

Figure 2/6: Single fold and multifold seismic section comparison. On the single fold section pronounced multiples interfere with reflections from a target horizon at between 1.05 and 1.15s two-way time.
(*Courtesy: Mobil Oil Canada Ltd*).

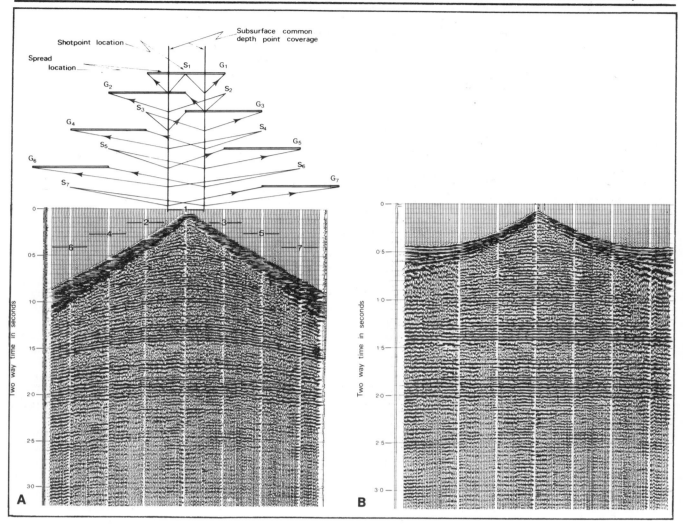

Figure 2/5: Expanded spread shooting for velocity determination and identification of multiples. **a)** The same zone of common depth point (CDP coverage) is tested seven times by shooting at locations S_1 to S_7 the geophone spreads G_1 to G_7 located with varying offsets. **b)** Application of normal moveout and datum reflection flattening makes identification of multiples easier because of their residual moveout.
(*Courtesy: Mobil Oil Canada Ltd*).

Suppression of direct noise waves is attained by using an appropriate number of detectors, spaced areally or linearly at pre-determined intervals. The problem of surface noise is greater in land surveys than marine surveys and usually it is necessary to conduct noise spread or 'walkaway' surveys (see figure 2/7) in each new area using single phones per trace over the total length of the planned spread. Once the period T of noise has been established it is possible to plot an array response graph (figure 2/8) which indicates the attenuation achieved by a particular array within a particular wavelength band. If λ_L and λ_S are respectively the longest and shortest coherent noise wavelengths it is desired to suppress, then simple design criteria indicate that the number of geophones N should be greater than $2\lambda_L/\lambda_S$, the length of the array L equal to $1.5\lambda_L$ and the detector spacing $\triangle_L$ equal to $1.5\lambda_L/N$. In the above only a uniformly spaced linear array has been considered; non-uniformly spaced linear and areal arrays

are also in wide use. In practice, economics and operational considerations combined with the unpredictability, throughout a survey area, of the strength and characteristics of coherent noise, dictate that the selected design will be a compromise rather than ideal for each individual shot location. It should be noted here that the first direct arrivals, or first-breaks, provide valuable near surface refraction information, and in both land and marine work these are not suppressed in the field. Where they interfere with desired shallow reflections they are muted during the data processing stages as described in the following chapter.

Source arrays can be used for coherent noise suppression either as an alternative to geophone arrays, or more commonly, as a complement. The same principles of design as for detector arrays are involved and this is most important for proper utilisation of low energy surface sources such as dinoseis, vibroseis and thumper.

So far we have discussed design criteria in terms of the spacing of detectors, the number of groups (or sections in marine work) to be used and the level of fold or stacking multiplicity. A final consideration is that of the length of spread. Design criteria here are not so specific, but the length of spread used is usually related to the depth of geological objective; long spreads are used to investigate deep structure, short spreads to obtain highly resolved data on shallower objectives.

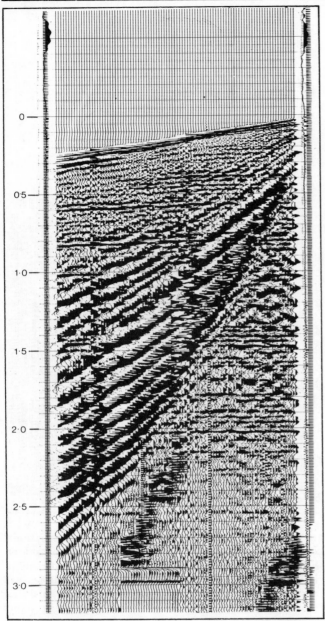

Figure 2/7: Example of a noise spread record. Each trace is derived from a single geophone. Spacing is usually 3—9m. Severe interference is seen between flat-lying (primary and multiple) events and sloping events (refractions, ground roll and air waves). (*Courtesy: Mobil Oil Canada Ltd*).

In marine work, the spread is towed in an 'end-on' configuration whereas on land it may be arranged end-on or either side of the source location (shot-point) as in 'split' or 'straddle' spreads. It should be noted, however, that with symmetrical split spreads the stacking multiplicity, which is defined as:

Number of channels x Interval between detector stations,

2 x Interval between source locations

while being fully effective in signal to random noise enhancement, it is however less effective than an ideal end-on spread in attenuating multiples. The duplication of common offset traces with symmetrical split spreads means that the records provide equal multiplicity (to end-on-spreads) as far as random noise attenuation is concerned,

but only half-fold multiplicity for multiple attenuation. In most situations, both on land and sea, it is common to leave a gap between the source and the first detector station (or even within a spread) to avoid interference from low velocity noise trains.

2.2.1 Grid design of a seismic programme

In an area where operational conditions and geological problems are well defined it is a relatively simple task to define the required layout of a seismic grid. Where shooting is to be undertaken in a new area, several factors need to be considered of which the more important are orientation and line-spacing of the seismic grid, field operation logistics, and budget.

Where operational conditions allow, it is generally advantageous to orientate a seismic grid such that lines are closely spaced parallel to the regional dip, or normal to the trend of major faults, anticlinal and synclinal axes, with only an open spacing of lines at approximately 90°.

In absence of information on the size of potential targets, definition of line-spacing will vary according to circumstances. Offshore, the high cost of exploration drilling and development suggest that only large scale structures are likely to be prospective. On the other hand, the costs of seismic surveys at sea are less costly than on land. As a consequence, it is common for marine investigations to commence with a survey of a widely spaced reconnaissance grid which is infilled over identified target structures at a later stage. If the size of potential target structures can be estimated, then the grid spacing should directly relate to this estimate. A spacing of half the width of such a structure would have a high probability of detecting its presence.

Field operational logistics will always have some influence on the layout of seismic programmes. Offshore, shallow water may prevent complete coverage of a prospect area, or it may necessitate splitting the programme into shallow and deep water surveys with use of appropriate vessels and equipment. Tidal conditions may influence orientation of lines to avoid excessive feathering of the streamer (the feather angle at any location is the angle between the line of the streamer and the line of profile and for good CDP stacking this is kept to a minimum except when using specialised acquisition techniques to obtain three dimensional coverage). On land, the distribution of access roads may influence the layout design, so might the existence of difficult terrain, bogs and marshes etc.

Budgetary constraints show marked contrast between onshore and offshore operations, both in respect of seismic survey and drilling costs. Offshore the cost of a 3000 m exploratory well may be as high as £1,000/metre whereas on land, a similar well might be drilled for £150/metre. Proprietary seismic data offshore may cost (inclusive of processing) approximately £200/kilometre, whereas on land costs are very variable depending on method used and local conditions, but they are always likely to be much higher. If a budget is severely limited then it will be more important to eliminate areas of low target probability. This can be done through a careful evaluation of known geology, and the interpretation of gravity, magnetic and shallow geophysical data prior to definition of the seismic programme layout (see chapter 7).

A case history involving reorientation of successive surveys as knowledge of first the regional dip, then the

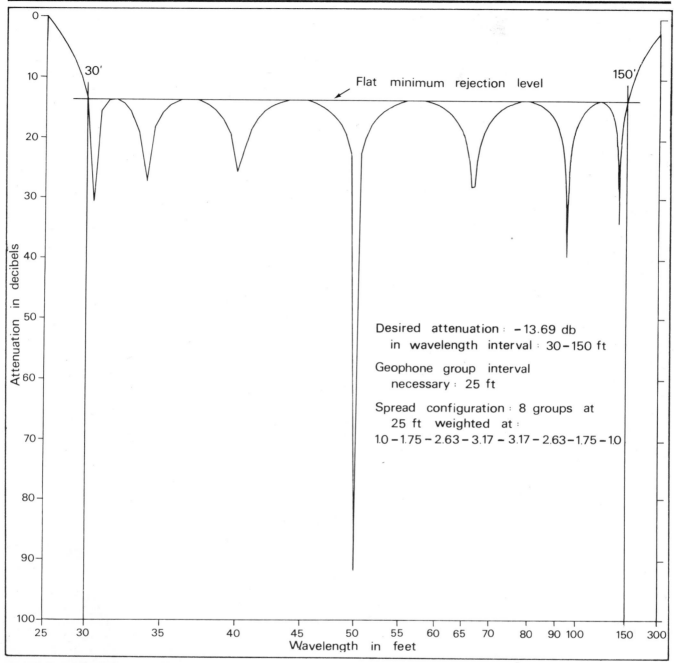

Figure 2/8: The response of a geophone spread designed to attenuate unwanted noise in the wavelength band 30ft to 150ft. (*Courtesy Digitech*).

strike of controlling faults became known is illustrated in figure 2/9.

2.2 The seismic pulse

At the beginning of this chapter we referred to the earth filter as one of the convolution factors of the seismic pulse which contributed to the final signal output. The earth's filter attenuates the seismic pulse in three ways: absorption of energy through conversion into heat; spherical divergence of the wave front; and reflectivity losses at acoustic interfaces. The attenuation involves amplitude and. frequency and these are interrelated. In normal seismic prospecting, the bandwidth of useful frequencies is extrem-

ely narrow and except for special shallow, high frequency input, high resolution surveys, it is limited to around 5–100Hz. Commonly, at exploration depths around 10,000ft or deeper, frequencies higher than 40Hz are rare. This means that thin bed resolution becomes increasingly difficult with increased depths of exploration. For a 40Hz wavelet, and an interval velocity of 15,000ft/s, beds thinner than 94ft, the quarter-wavelength one-way time width would be irresolvable. The band-limiting nature of the earth's filter and its frequency attenuation with depth is graphically illustrated in figure 2/10. At 12,000ft, the 40Hz amplitude factor of a waveform is one third of the 20Hz factor, while that for a 100Hz waveform is about one fortieth. It should be appreciated that the ideal seismic

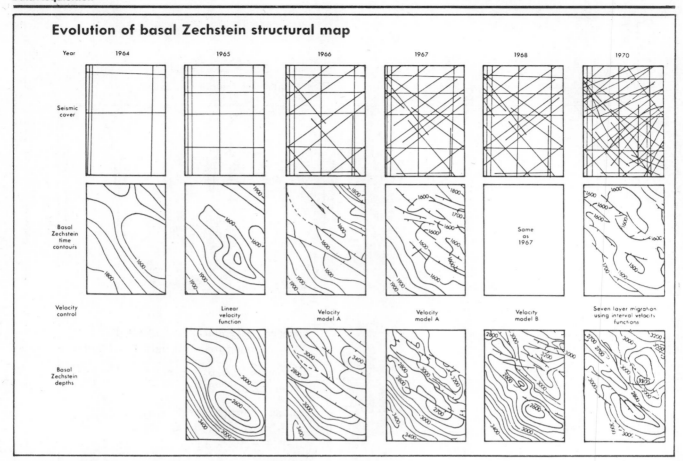

Evolution of basal Zechstein structural map

Figure 2/9: Seismic grid survey history from the North Sea, the West Sole gas field. Shooting dip lines in 1966 identified faulting, after which all follow-up surveying was reorientated to investigate the nature of these faults (after J.T. Homabrook. 'Seismic re-interpretation clarifies North Sea structure'. *Petroleum International*, April—May 1974).

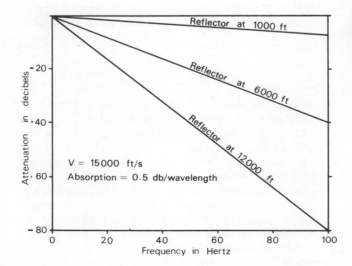

Figure 2/10: Reduction of seismic spectrum bandwidth with depth of reflector. High frequency events are severely attenuated at depth.

VVDA vibrator

1 Internal combustion engine
2 Adapting gear
3 Axial-flow pump (vibrator unit)
4 Lift system
5 Vibrator control unit
6 Vibrator
7 Hydraulic accumulators
8 Oil tank (main oil circuit)
9 Oil cooler

pulse from an implosive or explosive source should be as close to a spike as possible, which infers an instantaneous build-up of energy with a rapid decay and minimum reverberation or development of source-generated interference. In practice this ideal is impossible to attain, but the closer it can be approached the greater the high-frequency content of the recorded signal.

2.3 Land sources
2.3.1 Dynamite
Dynamite is the traditional material used to generate pulses and in 1975 it was utilised in over one half of worldwide surveys. Normally, charges tailored to the required depth of penetration are detonated in bedrock beneath the highly absorptive weathered rock layer, and in a hole which is tamped with water or mud above the charge. The quality and quantity of the input acoustic energy is related to the size and shape of charge and the depth at which it is exploded as well as to the type of surrounding rock material. Experimentation is required in each new area and in poor record areas, multiple charges may be required.

2.3.2 Vibroseis*
This mechanical wave-generator method is rapidly increasing its share of the acquisition market. In 1975, thirty percent of worldwide surveys employed this source and in built-up Europe the proportion was fifty percent. Its growing popularity is undoubtedly due to its inherent non-infringement of safety and ecological standards. A truck-mounted vibrator (figure 2/11) is coupled to the land surface and a long train of waves of progressively varying frequency is generated over a period of around seven seconds. The outputs from either an upsweep (increasing frequency input) or a downsweep (decreasing frequency input) are summed and correlated with the input sweeps to provide a 'conventional' field record (figure 2/12). Usually several vibrators sweep simultaneously in source arrays appropriately designed to attenuate surface noise and to improve the weak signal to noise ratio intrinsic to land-surface sources. Static corrections are obtained by shooting up-hole surveys and short refraction lines.

*   Trademark of Continental Oil Co.

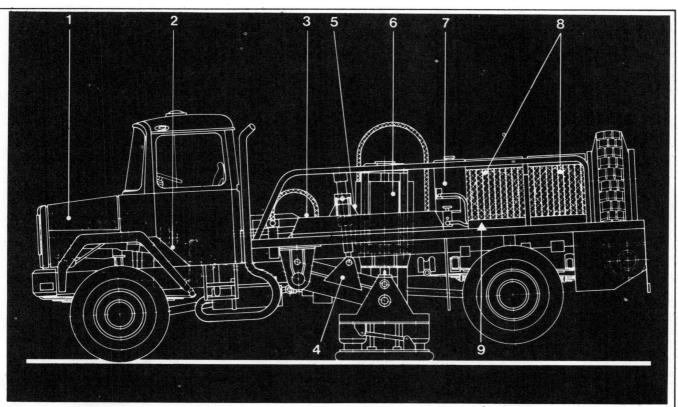

Side view of the vibrator system VVDA

Figure 2/11: Truck mounted vibrators used as seismic source in vibroseis surveying.
(*Courtesy: Prakla—Seismos*).

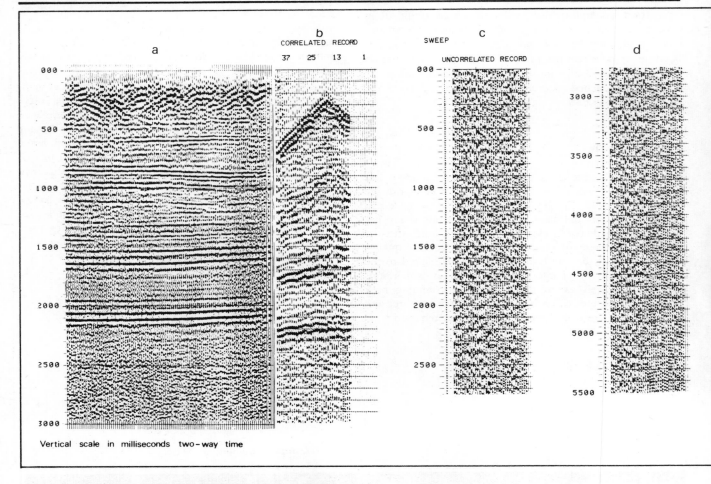

Vertical scale in milliseconds two-way time

Figure 2/12: Vibroseis: **a)** 2 x 6 fold stack, **b)** correlation corrected record; **c)** to **h)** sweep and uncorrelated record traces for time intervals 0.0–2.75s, 2.75–5.5s . . . . . 13.75–16.0s. Horizontal scale 30 traces/km. At this shot-point, sixteen 12s downsweeps ranging from 56–10Hz were applied. The 2 x 6 fold stack shows good quality down to economic basement at approximately 2.10s. (*Courtesy: Grant Geophysical*).

2.3.3 Dinoseis
Like Vibroseis, this source has advantages over dynamite sources. Basically it is a gas exploding device with the energy being imparted to the land surface via a heavy plate rammed downwards by the force of the explosion. Like Vibroseis, the signal to noise ratio is improved by the use of several units and/or the recording and stacking of several 'pops' at each station. Again it is necessary to run a supporting weathering correction survey.

2.3.4 Thumper or weight-dropping
This surface device is not in widespread use but is the leading seismic energy source in the desert areas of North Africa and the Middle East. Typically a 2–3 ton plate is dropped about 10ft. Summing of several drops is required and weathering correction data are collected separately as for Dinoseis and Vibroseis.

2.3.5 Air guns
This type of source, originally developed for marine surveys (see below) has been modified for use on land and is being increasingly used for work either in fresh water surveys or on dry land where they can be exploded in a suitably contrived liquid environment.

2.3.6 Mini-Sosie seismic source
A new method currently becoming operational provides a variation on the Vibroseis theme; a modified manually operated pneumatic hammer (figure 2/13) vibrates a plate coupled with the ground in a non-controlled random sequence. The signature of the random source energy is recorded by a sensor on the vibrating plate and is correlated with the recorded output to provide a conventional record. Again, summing of several signals is required. Indeed, the source is capable of producing 600 pops/minute and the methods depend on the summing of the results of a few hundred pops for each shot-point location. It is estimated that 8 to 12 records per hour can be obtained during smooth operations allowing approximately one mile of seismic survey per production day per seismic team. Use of this system is confined to shallow, high resolution seismic investigations more generally associated with engineering and mining studies than with hydrocarbon exploration. In figure 2/14 is shown a comparison of a Mini-Sosie record with a dynamite record, and in figure 2/15 a section obtained during a coal exploration programme in which good resolution of near surface faulting is clearly evident.

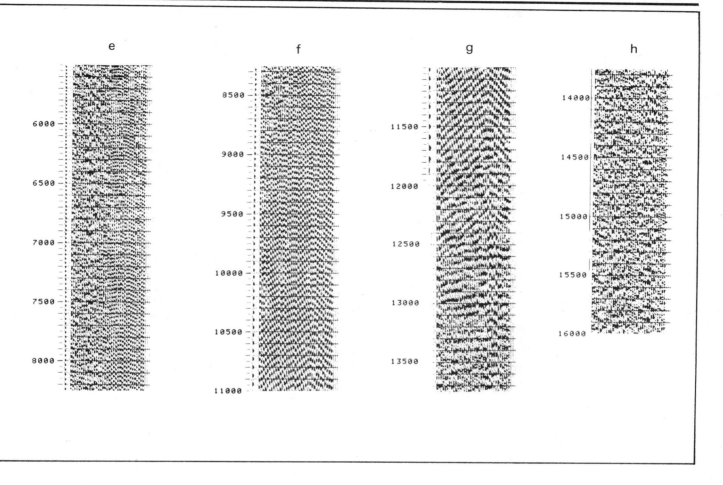

e   f   g   h

Figure 2/13: Mini-Sosie seismic source. A signature recording sensor is mounted on the top side of the vibrator base-plate. (*Courtesy: Société Nationale Elf Aquitaine (Production); SNEA(P).*).

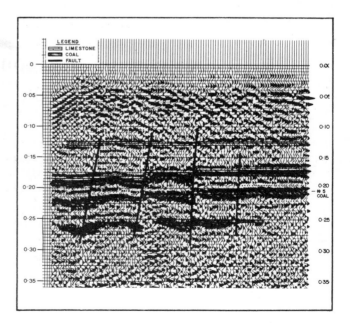

Figure 2/15: Mini-Sosie record showing high-resolution of faulting which affects a group of coal seams. (*Courtesy: Geoterrex*).

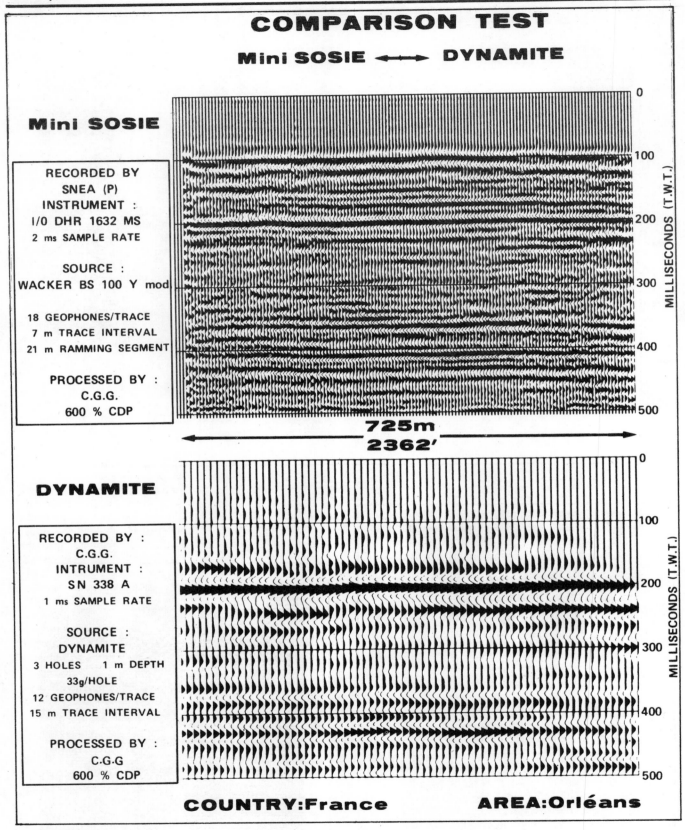

Figure 2/14: Comparison test between use of Mini-Sosie and dynamite as a seismic source.
(*Courtesy: SNEA(P).*).

2.4 Marine seismic sources

Well over twice as many miles of marine seismic line were surveyed in 1975 as on land. Work at sea has provided in recent years substantial technological advances in both source and streamer design. The major limiting factor in obtaining an ideal seismic pulse from marine sources is due to the effect of what is known as the bubble pulse. In marine seismics, simple explosive sources generate an unwanted train of secondary pulses produced by the oscillation of a gaseous bubble in the water body. Little pressure is generated as the bubble expands, but as it implodes under hydrostatic pressure, water surrounding the bubble compresses and a pressure wave is radiated. Before condensation or break-up of the bubble, periodic oscillation will occur, the period T is given by the Rayleigh-Willis equation:

$$T = \frac{0.045Q^{1/3}}{(D+33)^{5/6}}$$

where Q is the energy in joules and D the depth in feet to the centre of the bubble. In figure 2/16 the comparison is shown between bubble oscillation periods for various seismic sources and the Rayleigh-Willis relationship. High energy sources which have larger bubble periods conform more closely to the Rayleigh-Willis curve, and in seismic processing, it is preferable that the period should be so large that there is no interference with the primary source signature. In such a case the bubble oscillation signature can be removed using specifically designed deconvolution operators.

Several solutions have evolved to combat the bubble pulse problem. With conventional explosives, the solution is to release the energy at such a shallow depth that the gas bubble is vented into the air. The results are spectacular, but grossly inefficient as a large proportion of the energy is wasted. Thus in early days of offshore shooting, it was necessary to utilise large explosive charges, typically 50 lb TNT, a costly operation of benefit principally to the explosive manufacturing companies. Such large explosive charges are now seldom used. In one system, Maxipulse (a Western Geophysical system) a small explosive charge is used, about half a pound in weight, and de-bubble processing is used to remove bubble effects. To do this it is necessary to embody a gun-channel record in the system as an auxiliary to the streamer records, and this is used to produce a filter operator which collapses the bubble pulse sequence into the equivalent of a simple explosion pulse. An alternative way of counteracting the bubble effect is to dissipate the bubble energy. This technique is used in the Flexotir (Institut Francais du Pétrole) method. A small charge is exploded within a perforated steel chamber. Following the initial explosion, flow of fluids through these perforations to a large extent attenuates the bubble pulse. Implosion sources have been developed, for example Vaporchoc (Compagnie Générale de Géophysique), which do not have a bubble effect. With Vaporchoc steam is injected into the water to form a bubble which condenses and implodes, bubble collapse provides the seismic pulse and there are no secondary impulses. Systems have also been developed which generate explosions within a flexible container and vent the exhaust gases through hoses to the atmosphere. Conventional airguns (see p. 23) have been modified so that air is bled into the air bubble to increase pressure during collapse and inhibit oscillation; such a

modification is termed a wave-shaping kit. Perhaps, however, the most widely used method of countering the effects of bubble oscillation is to design a tuned array of sources, particularly when air guns are used. Two techniques can be used. Firstly, a number of sources may be fired simultaneously with different energy levels and hence different bubble periods so that the summation of the wavelets produces the desired source signature. Secondly, a number of sources may be fired with pre-set delays. With proper design, in either case, the end result is destructive interference of the bubble pulses.

The 1977 Society of Exploration Geophysicists' annual survey of world-wide geophysical activity shows that air guns are by far the most favoured marine source, these being used in 76 per cent of all data acquisition. Implosive devices followed with a 19 per cent share and gas/sleeve exploders had a 3 per cent share. The increasing popularity of air guns (and to a lesser extent, implosive devices) is provided by the magazine *Offshore* (January 1976) where of a total of 77 vessels working world-wide then, data on 66 were specified as using the following sources.

Air gun arrays	34
Gas/sleeve	18
Maxi-pulse	7
Vaporchoc	5
Single air gun	2

Two of the reported 14 contractors operated over one half of the vessels: Western Geophysical were the largest operator with 22 and Geophysical Service Inc had 17 vessels.

2.4.1 Air guns

An individual air gun generally consists of a system of two high pressure chambers (see figure 2/17), connected and sealed by a double-ended piston. During the charging cycle, air at high pressure (say 2,000 psi) is fed into the upper chamber and bleeds through the hollow piston into the lower chamber. The piston assembly is held in the downward position because the area of the trigger piston is larger than that of the firing piston. To fire the gun, an electrical pulse opens the solenoid valve and a slug of high pressure air is delivered to the underside of the trigger piston. The piston shoots upwards under the pressure exerted on the firing piston releasing the air in the lower chamber into the water. Pressure in the upper chamber then drives the piston back to its initial position and the charging cycle recommences. During a period of a few milliseconds, all the high pressure air in the lower chamber is vented into the surrounding water through centrally located ports and it is the explosive release of this air which provides energy for the initial seismic pulse.

Guns range in size from a few cubic inches (lower chamber capacity) to about two thousand cubic inches, and operate at pressures 2,000-3,000 psi. As described earlier, multiple arrays of guns are usually employed, towed behind or alongside the survey ship on a suitable frame at depths of about 10m. The differences in pulse shape generated by a single air gun, a wave-shaped air gun and an array of air guns is illustrated in figure 2/19.

2.4.2 Sleeve exploders

The sleeve exploder, also called Aquapulse, is a seismic source developed by Esso Production Research, and it is

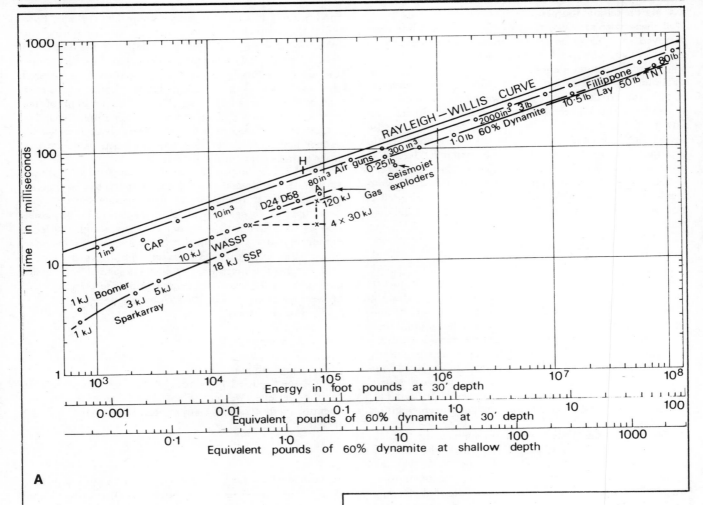

A

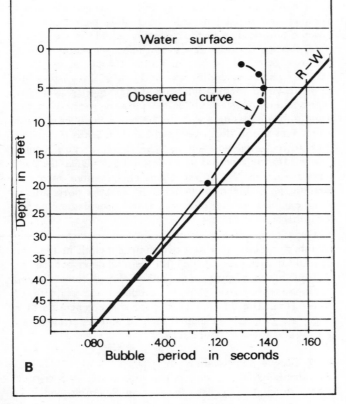

B

Figure 2/16: a) Bubble oscillation periods for various seismic sources compared with the Rayleigh-Willis curve. **b)** Plot of an observed bubble oscillation period versus depth; 300in^3 at 2000 psi of air. R—W is the Rayleigh-Willis curve. Near surface, the bubble period shows a divergent decrease due to the mass unloading effect which occurs as the bubble approaches the air/water interface.

used under licence by several contractors and has other alternative trade names. In figure 2/18 an Aquapulse system is shown as used by Western Geophysical. Propane and oxygen are separately fed into the rubber sleeve and ignited by a sparking plug. The high energy pulse is generated as the sleeve expands; after expansion, the sleeve contracts with venting of gases to the surface, decompression being assisted by cooling and pumping. Multiple arrays such as shown in figure 2/18 are generally required to provide sufficient total energy output for deep penetration.

2.4.3 Maxipulse

Maxipulse is one of the simplest seismic sources currently in use. A small cylindrical charge of nitrocarbonitrate explosive is projected by water pressure down a hose to a submerged gun, striking its detonator on a firing wheel as it is ejected, see figure 2/20. A percussion cap is activated and this fires a delay fuse which detonates the main charge after sufficient period for the gun to have been towed a few metres from the explosion point.

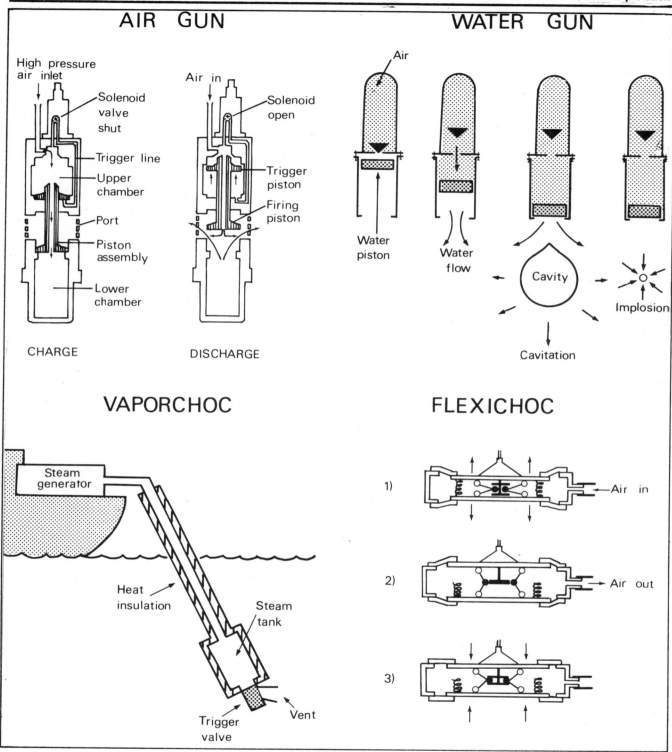

AIR GUN

High pressure air inlet

Solenoid valve shut

Trigger line

Upper chamber

Port

Piston assembly

Lower chamber

CHARGE

Air in

Solenoid open

Trigger piston

Firing piston

DISCHARGE

WATER GUN

Air

Water piston

Water flow

Cavity

Cavitation

Implosion

VAPORCHOC

Steam generator

Heat insulation

Steam tank

Trigger valve

Vent

FLEXICHOC

1)

Air in

2)

Air out

3)

Figure 2/17: Schematic diagram showing the principle of operation of four different non-explosive seismic sources. (After McQuillin and Ardus, 1977).
Top left: PAR Air Gun (registered trade mark of Bolt Associates. The air gun is patented under British Patent Specification Nos. 1,090,363; 1,175,853; and 1,322,927).
Top right: MICA Water Gun.
(Courtesty Société pour le Developpement de la Recherche Appliquée (SODERA).)
Bottom left: VAPORCHOC seismic source.
(Courtesy Compagnie Générale de Géophysique (CGG).)
Bottom right: FLEXICHOC seismic source.
(Courtesy: Géomécanique).

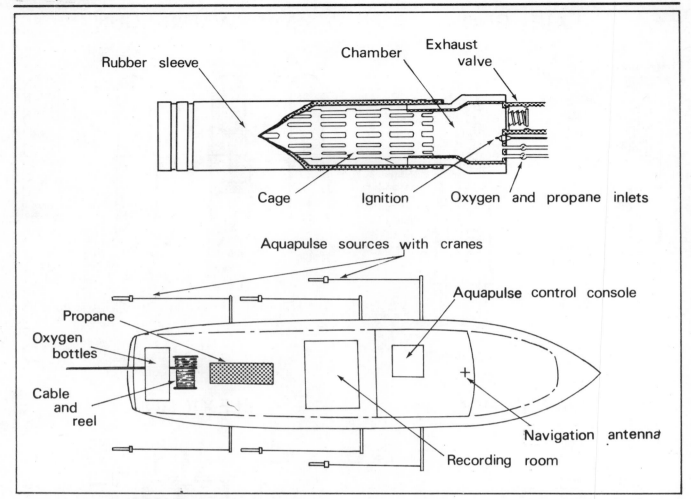

Rubber sleeve

Chamber

Exhaust valve

Cage

Ignition

Oxygen and propane inlets

Aquapulse sources with cranes

Aquapulse control console

Propane

Oxygen bottles

Cable and reel

Navigation antenna

Recording room

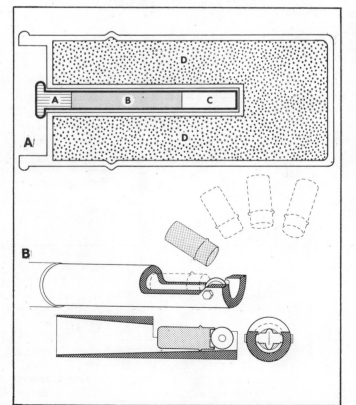

A

B

C

D

D

A

B

Figure 2/18: Schematic diagram (above) of an Aquapulse seismic source with (below) a ship layout showing a towing arrangement for the deployment of an array of six sources. (*Courtesy: Western Geophysical*).

Figure 2/20: a) Schematic cross section of Superseis charge used in the Maxipulse system. A — rim-fire percussion cap, B — delay column, C — booster and D — nitrocarbonitrate main charge. b) Maxipulse gun. Top — perspective view of gun showing charge strike the firing wheel, then successive stages of ejection. Bottom — cross section and end views of charge in gun at instant of striking the firing wheel. (*Courtesy: Western Geophysical*).

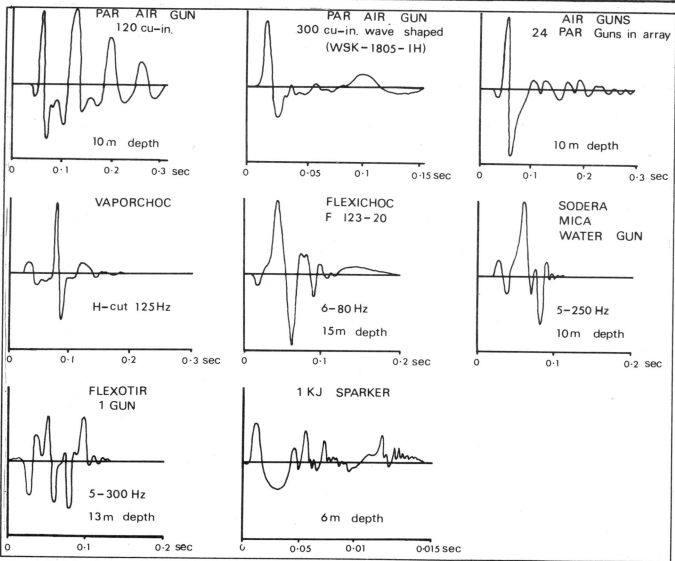

Figure 2/19: Signatures of a range of seismic sources.

The firing gun assembly has a gun transducer attached to it which is used to record the source signature. As described above, no attempt is made to attenuate the bubble pulse in the acquisition stage; this is adequately performed by specialised pre-processing of the data.

2.4.4 Vaporchoc

Vaporchoc (used by Compagnie Générale de Géophysique) is an implosion source fuelled by superheated steam (see figure 2/17). A steam generator is located on board ship and steam is fed into the water through a remotely controlled valve to form a steam bubble. When the valve closes, the steam in the bubble condenses and the bubble implodes under hydrostatic pressure. During implosion, high pressure develops in the water around the wall of the diminishing bubble and acoustic energy is radiated as a primary impulse with negligible secondary oscillation on complete collapse of the bubble (see figure 2/19).

2.4.5 Flexotir

The Flexotir source was developed by Institut Français du Pétrole. A small explosive charge is flushed down a hose into a submerged perforated steel cage of diameter slightly larger than that attained by the bubble at maximum expansion. The charge is fired electrically. Oscillation of the bubble is inhibited and the bubble pulse attenuated by the flow of fluids in and out of the perforations. As seen in figure 2/18, the signature of a single gun is quite complex, but as with air guns, improvements can be attained by operation of an array of sources.

2.4.6 Water guns

The water gun is a variety of air gun which has been designed by SODERA (Société pour le Developpement de la Recherche Appliquée) as an implosive seismic source which generates an intrinsically clean signature (see figure 2/19). On firing, the compressed air in the gun chamber rams a piston forward ejecting a volume of water through a number of ports. When the piston is arrested, cavitation occurs and the main acoustic energy is generated by implosion of the water body through cavity collapse. No bubble pulse is produced; the air in the chamber is vented and the charging cycle recommences. One advantage of this source is that its features allow flexibility in designing source arrays which can be effective in attenuating coherent noise.

2.4.7 Sparker systems
In sparker source systems, acoustic energy is generated by electrical discharges in sea water. Generators are used to charge capacitor banks which can then be triggered to discharge high voltage (3–10kV) through spark tip arrays towed in the water. Low energy sparkers, 100 joule–5 kilojoule, are widely used in single channel seismic profiling as part of shallow geological studies, engineering site surveys etc, (see McQuillin and Ardus, 1977). High energy sparkers, energy up to 200 kilojoule, are used as a source for conventional seismic work, one advantage being the relatively low operational cost of using this type of system. In recent years, a more significant use of the sparker as a source for multichannel seismic acquisition is in the field of shallow gas detection, in particular for drill site surveys. Here sparkers are operated in the 3–5 kilojoule range and data acquisition aims for high resolution data in the 0–1s two-way reflection time range. Sections are processed to give true amplitude recovery (see chapter 3) display as well as conventional displays.

2.5 Geophones and hydrophones
In seismic surveying, two types of acoustic detecting transducers are used, geophones on land and hydrophones in marine conditions and in mud-filled boreholes. We have seen in chapter 1 that in reflection surveying we are attempting to record trains of P-waves, or compressional waves as they pass a specified point. Thus pressure sensitive phones are ideally suited for marine work, recording as they do the pressure changes above or below ambient water pressure. On land, pressure-sensitive phones cannot be used as it is generally impractical to bury the phones in such a way that they would have adequate fluid coupling with surrounding material. The use of pressure sensitive phones at or near ground surface is further complicated by the fact that the air to ground interference acts as a phase-change reflection boundary at which differential pressure approaches zero. For these reasons land geophones are of the type sensitive to particle motion.

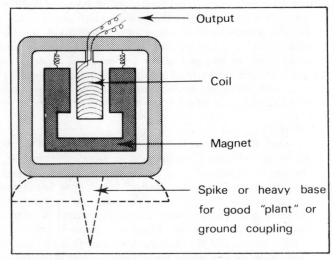

Figure 2/21: Schematic of a velocity, or particle motion geophone. In this type the coil is fixed and the magnet suspended; alternatively, the coil may be suspended and the magnet rigidly fixed to the geophone case.

2.5.1 Land geophones
Figure 2/21 illustrates the principle of a geophone designed to measure particle motion by conversion to electrical energy. Arrival of a compressional wave sets the earth's surface in motion; the geophone case with rigidly attached coil is coupled to the ground and moves in sympathy whereas the magnet which is suspended on springs remains effectively stationary because of its inertia. Movement of the coil within the magnetic field induces an electrical voltage across the coil which is proportional to the velocity of the coil with respect to the magnet. An alternative, and more common design is to mount the magnet rigidly and suspend the coil around an inertial mass; the principle is identical and such geophones are called moving coil geophones. In both cases the output is independent of frequency at frequencies above the natural frequency of the suspended element; below natural frequency the response is frequency dependent. Modern geophones use a dual coil system, the dual, series-connected coils reduce external interference.

In land surveys, it is important that good coupling is obtained between the ground surface and the geophones. The cases are usually either heavy with a flat base, or light with a coupling spike which is pushed into the soil. Geophones are usually deployed in groups at each detector location, the phones being connected in series. The group is called a string, and each string is attached to a takeout in the main cable which feeds to a specific channel in the recording system. For speed and efficiency during a survey of a seismic line, several cables are laid out so that the number of strings deployed is greater than the number of channels being recorded. After each seismic shot, or a fixed number of shots, a string or a number of strings are picked up from the end of line already surveyed and 'leap-frogged' to the other end. The recording system incorporates a 'roll-along' switch which is used to drop-out and pick-up the appropriate cable take-outs.

2.5.2 Hydrophones
The principle of the marine hydrophone is very simple. Within the phone a piezoelectric transducer produces voltages in response to pressure changes caused by the passage through surrounding water of seismic pressure waves. Figure 2/22 shows the frequency response of a typical hydrophone. For static cable recording, good response can be obtained with a simple hydrophone element, but current methods employ continuous profiling using towed streamers and substantial noise is generated through vibration of the cable (strumming) and sudden acceleration/deceleration effects produced by heave acting on the towing vessel, the vessels movements being transmitted to the streamer. Various methods are employed to reduce these effects:

1. Ship motion is decoupled from the streamer by using an elastic non-active lead-in section; this absorbs the ship's heave motion allowing the cable to be towed at a constant speed through the water.

2. Streamer depth controllers or 'birds' are used to maintain constant streamer depth along the length of the streamer. Each controller clamps to the cable and has fins which are servo-linked to either pre-set pressure sensors, or pressure

sensors which can be adjusted by remote control. Typically about 6 to 8 controllers might be attached to a long marine streamer.

3. Lead-in sections to the cable can be faired to reduce noise induced by strumming.

4. Instead of single crystal element hydrophones, dual crystal, acceleration-cancelling phones are used which have a very low sensitivity to horizontal accelerations, one of the main sources of noise problems.

The acceleration-cancelling hydrophone is an important recent innovation. A schematic diagram of one type is shown in figure 2/23. Each crystal element consists of an annular piezoelectric ring, metallic-coated on both surfaces and bonded at the open ends by thin convex metallic diaphragms. In the towed streamer configuration, vertical pressure fields produce enhanced responses due to the pressure vectors generated in the annular rings by the diaphragms. Furthermore, axial accelerations generate equal but opposite phase signals in the twin elements in each geophone which when summed effectively cancel each other out. Among the benefits of using streamers incorporating such hydrophones, as well as other noise reducing design criteria, are firstly that marine surveys can now (as opposed to a few years ago) be conducted at higher tow speeds and in rougher weather conditions, and secondly, improved signal to noise ratio allows the geophysicist to obtain deeper information and better structure delineation after data processing.

2.6 Recording equipment
The field equipment required to record reflected seismic signals as detected by arrays of geophones consists basically of four units, each of which performs several functions (figure 2/24). A typical system would be as follows:

Recording amplifier, with
  a) pre-amplifiers,
  b) analogue filters including high and low frequency cut-off and anti-alias.
  c) multiplexer,
  d) gain-ranging amplifier,
  e) analogue to digital convertor.
Control or logic unit, with
  a) time break amplifier,
  b) formatter,
  c) digital gain control,
  d) digital to analogue converter,
  e) demultiplexer,
  f) playback amplifier,
  g) playback filters,
  h) total system controls.
Tape transport unit.
Camera unit.

A total field system would also include power supply, seismic channel and system test circuitry, test signal generator and system fault location units. Present day field equipment is contained in compact, lightweight modules which are capable of being back-packed for operators in difficult terrains, or built into container cabins, small vehicles etc for more normal operations. One such system is illustrated in figure 2/25.

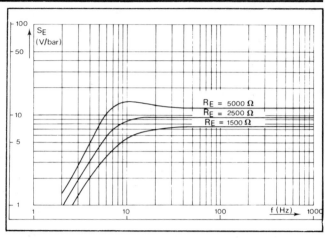

Figure 2/22: Pressure sensitivity versus frequency response curve for a Multidyne hydrophone group in a Prakla-Seismos Piezo Oil Streamer. Pressure Sensitivity S_E is related to the termination resistance R_E across the input to a DFS recording system. (*Courtesy: Prakla–Seismos*).

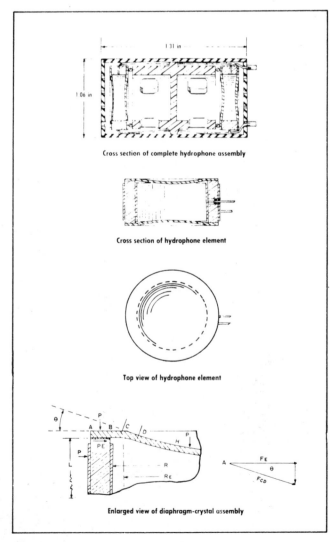

Cross section of complete hydrophone assembly

Cross section of hydrophone element

Top view of hydrophone element

Enlarged view of diaphragm-crystal assembly

Figure 2/23: Schematic of a Multidyne acceleration cancelling hydrophone. (*Courtesy: Seismic Engineering Company*).

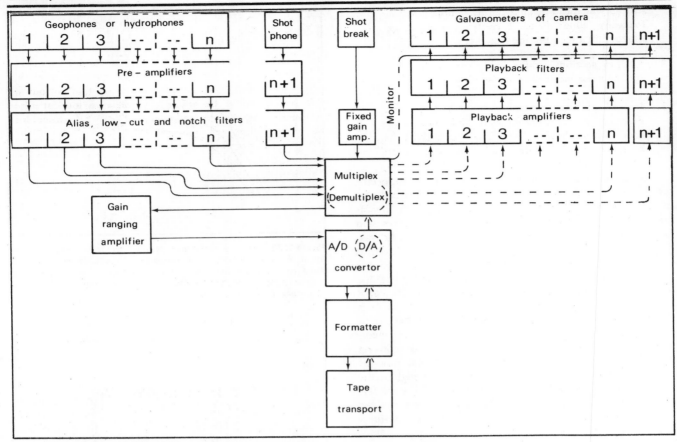

Figure 2/24: The principal components of a digital seismic recording system.

Figure 2/25: Texas Instruments DFS V seismic recording system. (*Photo: Techmation*).

Although analogue field acquisition systems are still in use in research and academic studies, most seismic exploration is almost exclusively conducted utilising digital recording techniques. Only digital techniques will be described here, information on analogue methods may be obtained by reference to one of the geophysical text books listed as suggested reading.

Digital field systems are designed to record seismic signals in digital form by sampling amplitudes at appropriate time intervals such that reversal of the process, digital to analogue conversion, will generate an output signal which compares with the original without loss of fidelity. The process is illustrated in figure 2/26. In the example shown a 62.5Hz wave (dashed line) is sampled every 2msec. On play-back, conversion to an analogue electrical signal produces a signal which may be conceived of as the original signal (smooth line) reduced to 97 per cent amplitude plus a ripple of higher order harmonics. Reproduction of the original signal is achieved by filtering with an analogue high-cut filter, thus removing the unwanted high frequency harmonics.

2.6.1 Recording amplifier

The functions of the various components of this part of the system are outlined as follows:

(a) Pre-amplifiers. These provide a means of selecting fixed gains for application to the input signal.

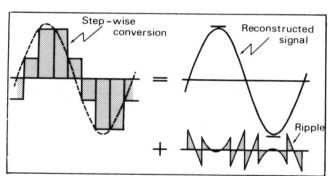

Figure 2/26: Digital sampling and playback (*Adapted from illustration in Pictorial Digital Atlas. Courtesy: United Geophysical Co.*).

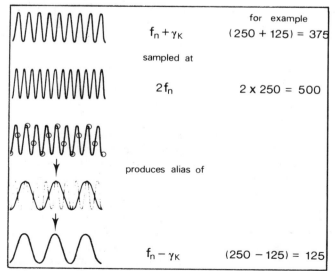

Figure 2/27: The aliasing effect of sampling a signal at too low a sampling frequency.

(b) Analogue filters. In land work, low-cut filters are tailored to the natural frequency response of the geophones being used. The cut-off is chosen above the natural frequency to avoid harmonic distortion and is generally in the range 7–14Hz. In offshore surveys, the flat frequency response of hydrophones means that an 'out' or 0Hz setting could be applied, however, a low-cut filter is usually applied to reduce streamer and other low frequency environmental noise.

High-cut filters must be applied with cut-off being related to the digital sampling interval being used. If the seismic signal is insufficiently sampled, aliasing may occur, and on playback artificial low frequencies may appear. The frequency below which aliasing does not occur is termed the Nyquist frequency and this is one half the sampling frequency that is, for 2msec sampling the Nyquist frequency would be 250Hz. If f_n is the Nyquist frequency and a signal of higher frequency $(f_n + \gamma_k)$ is sampled at the interval $1/2f_n$, aliasing produces an output signal of frequency $(f_n - \gamma_k)$. This is illustrated in figure 2/27. A signal of 375Hz is sampled at 2msec (500Hz) to produce an aliased signal of 125Hz; thus for 2msec sampling it can be seen that a high-cut anti-alias filter is necessary removing all signals of frequency higher than 250Hz. Figure 2/28 shows a suite of filter amplitude response curves which might be used for 2msec sampling acquisition.

The third type of analogue filter which is sometimes used is a notch filter. This type of filter is used for example to eliminate high-voltage power line interference which is often a serious problem in land operations. Figure 2/29 shows the amplitude response curve for one such filter.

(c) The multiplexer. The previous steps of amplification are performed simultaneously but separately on each seismic channel. It would however be a most inefficient and cumbersome method of digitally recording the data, if each channel were to be separately recorded on tape. The function of the multiplexer is to switch through all data channels in sequence, allowing enough time (a few microseconds) for each signal sampled to charge a capacitor to the correct signal voltage level. Thus at each sample time, for example every 2msec, every channel is sampled and the voltage level stored ready for output as a single data channel to further stages of amplification and analogue to digital conversion.

(d) Gain-ranging amplifier. This part of the system is at the heart of modern seismic instrumentation and on it depends the high dynamic range and fidelity of the recording process. In analogue recording, gain control is applied directly to boost or attenuate the input signal to keep it within the system's usually stringent amplitude limits. Several methods can be applied, but they all suffer from the same drawback; the signal is irretrievably altered and the process cannot be reversed unless a sophisticated method of recorded programmed gain is employed.

The principle of the digital gain amplifier hinges on its ability to simultaneously apply to the signal, and record separately, a required value of variable gain. Thus, by reversing the procedure, the true amplitude can be recovered on playback, and this is particularly valuable in detailed stratigraphic and hydrocarbon studies. Current limitations on true amplitude recovery display are confined to those associated with available display methods. The process by which the amplifier operates is comparative. As

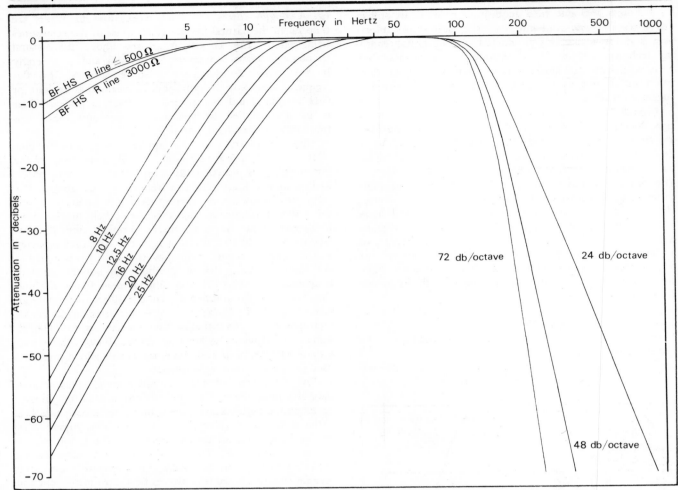

Figure 2/28: Response curves for a typical range of high and low cut-off options available for filtering input signals to a seismic recording system. This set is designed for use with a 2msec sampling frequency.
(*Courtesy: Sercel*).

described above, after multiplexing, each portion of the sampled signal is stored in a capacitor and successively compared with reference voltage outputs. Current amplifiers are of either the binary or quaternary floating point types, ie they involve 2^1, 6dB gain steps or $2^2 (=4)$, 12dB gain steps. As an example the method used in a Sercel SN338B system is illustrated in figure 2/30. At the beginning of the sample processing cycle, the amplifier is set at the intermediate gain of 2^8. The output is fed to two comparators with two references 12dB apart. If the voltage falls within the bracket of the two voltages, the gate is unchanged. If the output voltage falls outside this bracket, the control logic opens another gate so as to increase or decrease the gain. After three such comparisons, the amplifier has reached the proper gain as can be seen in the schematic of figure 2/30. The time required to make these comparisons and decisions is 10µsec. After the third decision is made a conversion order is transmitted to the analogue to digital convertor.

The gain is thus established in the range 2^0 to 2^{14}, or 84dB.

(e) Analogue to digital convertor. After each sample has had the required gain established and applied, the resultant amplifier output sample is converted into a binary code word, generally of 15 bits, made up of sign plus fourteen

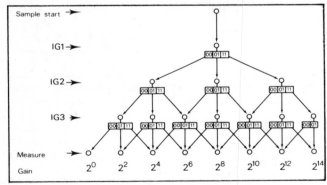

Figure 2/30: Comparator system for applying digitally recorded gain to a seismic signal.
(*Courtesy: Sercel*).

0 or 1 binary digits.

It is appropriate here to consider the intrinsic signal to noise ratio of contemporary instrumentation. The maximum dynamic range of reflection signals as measured at detector output is of the order 100dB above natural background noise. If we include potentially recoverable signals below ambient level, an additional 20dB can be included. Recorder and amplifier design thus aims for noise to be held down 30dB below ambient level, and this is generally attained. The desired 120dB signal range requires, therefore, a 130dB dynamic range of the recorder and this is adequately attained by an 84dB gain range amplifier linked to an analogue to digital signal convertor of the type described above.

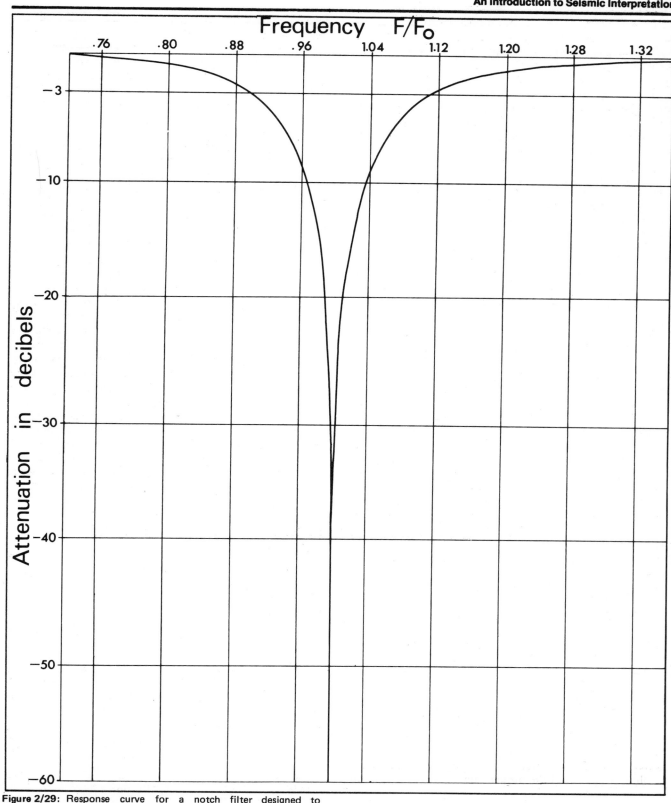

Figure 2/29: Response curve for a notch filter designed to attenuate power-line interference. (*Courtesy: Sercel*).

2.6.2 Control or logic unit

The functions of this part of the system are as follows:

(*a*) *Time break amplifier*. This handles the input from the time break (shot initiation) signal with output to a separate channel to the seismic signal channels. In figure 2/24 it is defined as the shot break fixed gain amplifier.

(*b*) *Formatter*. This instrument organises the digital word output from the analogue to digital convertor (A/D convertor in figure 2/24) ready for writing on tape. Current industry accepted formats are 9 track SEG B or C, although

21 track formats are also in use. Header information is fed to the tape in advance of the data record.

(c) *Digital gain control.* For playback purposes through the monitor camera, the recorded signals must be adjusted to fall within the dynamic range of the galvanometer of the camera. Automatic gain control is generally used.

(d) *Digital to analogue (D/A) convertor.* This converts back the digital record into analogue form for display on the monitor camera. Because of the limited dynamic range required, the most significant bits of the digital record are processed.

(e) *Demultiplexer.* Again, to allow display on the monitor camera it is necessary to reverse the multiplexing process through a demultiplexer.

(f) *Playback amplifier.* This adjusts the amplitudes of the output channels simultaneously prior to filtering.

(g) *Playback filters.* These provide a range of analogue filter settings, high and low-cut, which are adjusted to give the best quality monitor camera record.

(h) *Total system controls.* This unit includes plug-in connections to all other instruments and indirectly to array inputs etc. The geophysical observer has all the necessary switches and controls to operate the complete shot-firing and signal recording process from start to finish.

2.6.3 Tape transport unit
Digital tape decks are used to record the output from the formatter onto either 9-track ½ inch tape or 21-track 1 inch tapes in standard SEG formats. Data is recorded at either 800 or 1600 bpi at speeds of between 10 and 120in/s.

2.6.4 Camera unit
The camera unit contains galvanometers which are driven by the output from playback channels. Wiggle trace records are produced for each channel on light-sensitive dry-write paper. Two types of read-after-write records are used. For monitoring purposes, the direct input signals can be displayed, or alternatively, as a means of obtaining a provisional seismic section, the data can be played back and displayed after application of gain control and filtering.

2.7 Shot-firing control
In marine surveys, seismic acquisition is a continous automated routine. Most survey vessels depend for navigation and position-fixing on either integrated satellite navigation systems, or on radio-navigation systems, usually linked to a computer and data logger which provides control of the shot-firing sequence. For CDP stacking it is important (see figure 2/1) that distances between shot-points are constant. The complex interfacing of all systems is well illustrated in figure 2/31.

In land surveys, the firing routine is manually controlled by the observer. For surface sources the observer simply presses a firing button to initiate the shot. On dynamite crews, the observer is connected to the shooter's shot-box both by telephone and electrically, and for safety, the shot initiation is controlled by the shooter.

2.8 Quality control
Quality control of a seismic survey is principally the responsibility of the seismic crew's party chief and observer. Modern equipment incorporates special fault finding and monitoring facilities which allow faults to be quickly dis-covered and, hopefully, remedied. On most surveys, the client, who is paying for the survey, will have a representative attached to the project, and it is the responsibility of the client's representative to assist in drawing up the detailed job specification and ensure that this is adhered to at all times except with his approval if modification or relaxation can be seen to be advantageous to the client.

2.9 Survey accuracy and position-fixing
Seismic lines on land need to be surveyed to obtain the positions of shot-points and the geophone spreads as well as the heights above or below datum of each of these positions. Conventional land survey methods can be used and the required accuracies are usually easily obtained. Prior to shooting, lines can be staked out and shot-points surveyed in. Survey data are then used in the application of static corrections. In general there is little likelihood that for any individual project, the positional accuracies will be so large as to cause any significant seismic mis-ties (see chapter 5) at line intersections. In remote areas, for which detailed topographic maps are not available, it is important if at all possible, either to link the survey grid to a permanent and accurately located bench mark or beacon (which may be some distance from the survey area), or to obtain accurate geodetic control from astronomical observations or by using a satellite receiver. A significant recent land-surveying innovation is the use of vehicle-mounted inertial navigation equipment, which, when properly referenced can provide a high degree of accuracy. When the interpreter is handling data from such remote and poorly mapped areas these may well be from different surveys of different vintages and possibly by different contractors. Under such circumstances it is necessary to check the consistency between topographic surveys. Adjustment to a single datum or reference point may be necessary before interpretation commences and comparison of line elevations at intersections can be used as a check on the consistency obtained.

Surveys at sea obviously cannot be related to existing topographic maps; in this environment radio position-fixing or integrated satellite navigation methods are the most widely used. Accuracies range from a few metres to tens or even hundreds of metres. Generally, the highest accuracy systems have limited range and can only be used in areas close to the coast or in the vicinity of fixed platforms and drilling rigs.

Radio-positioning systems obtain fixes by determining distance values between survey points and shore or platform based transmitter/receiver reference stations. At any fix-position distance values to two or more such stations are obtained at approximately the same time thus defining the location of the fix. A simple position-fixing pattern is illustrated in figure 2/32, and in this case we see the method applied to a high precision (better than 5m accuracy), limited area survey problem, the survey of an estuary. Shore stations are positioned at A, B and C. A transmitter on the survey vessel at D sends a coded radio signal to each of the stations in turn and this activates the transmitter which then sends a responding pulse. By measurings the elapsed time between transmission and reception of signal and dividing by the speed of propagation of radio waves a range to each shore-station can be measured. Motorola Mini-Ranger III and Decca Trisponder systems operate in this fashion. (For further reading see

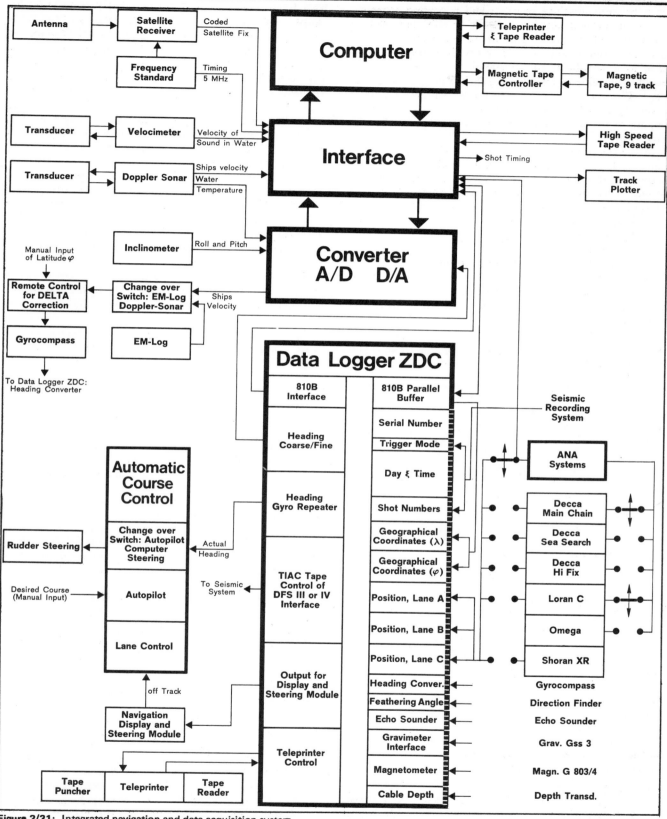

Figure 2/31: Integrated navigation and data acquisition system.
(*Courtesy: Prakla—Seismos*).

chapter 2 of McQuillin and Ardus 1977).

In comparing data from different surveys, the interpreter must again be aware of the possible uncertainties regarding position-fixing. The results of a single survey can show good internal consistency in terms of the fit of line intersections, but comparison of surveys using different position-fixing systems may indicate the need for adjustment, there being with all such navigation systems a difference between absolute accuracy and relocation accuracy. The accuracy with which a site may be relocated using an identical deployment of positioning equipment is higher than the absolute accuracy of any position-fix with reference to a worldwide co-ordinate system. It is sometimes possible to calibrate a positioning chain and define fixed (as opposed to random) errors which exist within the area of cover. Again, in comparing data from different surveys it is important to check which corrections have been applied for such errors.

Although radio-positioning systems such as Decca Hi-fix, Pulse 8, Loran-C, Toran etc are still widely used to give accuracies of 100m or better, an ever increasing proportion of seismic surveys are now located using integrated satellite navigation systems. A typical configuration might be:

> Satellite navigation receiver,
> Four-element doppler sonar,
> Gyro compass,
> Interface to radio-positioning systems.

Some systems are in use which employ integration of satellite and inertial navigation systems.

Satellite navigation depends on reception of signals on board ship from a group of satellites operated by the US Government; these are the Transit satellites and at the time of writing there are six in orbit all occupying polar circuit orbits circling the globe every 107 minutes, see figure 2/33.

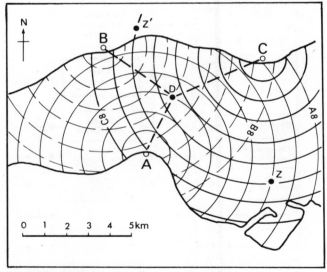

Figure 2/32: Range-range radio positon-fixing pattern in an estuary (after McQuillin and Ardus, 1977).

A satellite position is obtained only during periods when a satellite passes above the horizon as viewed from the survey vessel. Frequency of such occurrences vary with latitude, but good fixes every 1—3 hours are common. Signals are received through an omnidirectional aerial. These signals contain precise data on the satellite's orbit and the system also measures the doppler shift of the transmission of two carrier wave frequencies. A computer analyses this data to give the ship's position with reference to a satellite derived geodetic system. For accurate results the ship's speed and heading need to be well defined during the reception period of signals from the satellite, hence the need for integration with a gyro compass as well as either a radio-positioning system, or more commonly, a four-element doppler sonar.

This latter device gives an accurate measurement of the ship's speed over seabed by transmitting pulses of sound in four narrow beams as illustrated in figure 2/34. Comparison of the frequencies of the fore and aft received signals gives a measure of the doppler shift from which can be computed the vessel's along course speed. Comparison of the frequencies of the port and starboard received signals give similarly a measure of the ship's sideward drift. A computer is at the heart of the system and this is often used to control the shooting cycle during seismic surveys to give added precision to the location of shot-points for CDP stacking (see figure 2/31).

Contractors' reports or independent quality control reports usually contain estimates of survey accuracy. However this may vary within an area surveyed, in particular if both deep and shallow water areas are under investigation. Doppler sonar can only be operated with present-day equipment in water depths of up to between 200m and 500m, depending on equipment. In deeper water the signals from seabed are too weak and the sonar system tracks on layers within the body of seawater; in such circumstances positional data are generally of much lower accuracy. Integration with radio-navigation systems or an inertial navigator can obviate this difficulty. For best results, at least two good satellite passes should be recorded on each survey line. In surveys made up of a large number of short lines it can be costly to meet this condition and if positional uncertainty is suspected, lines surveyed in the absence of this control may need adjustment.

In most circumstances there should be no difficulty obtaining good line intersection ties between surveys made on the one hand with good radio navigation position-fixing and on the other with an integrated satellite navigation system. Nevertheless it should be noted that the values of latitude and longitude derived from Transit satellite data are in terms of a particular satellite derived world geodetic system and that it is necessary to apply corrections to such data to obtain positions in terms of other local or global systems such as the British National Grid or UTM. Often these corrections are applied by computer before data is output from the system but this may need to be checked if mis-ties occur between surveys.

Given that proper precautions have been taken, most modern marine data in water depths within sonar range or within the range of good radio navigation should have positional accuracy better than 50—100m and any necessary adjustment between different surveys should not greatly exceed this figure. High precision surveys using short-range radio location can be surveyed to an accuracy of approximately 5—10m.

2.10 Map scales and projections

Often the results of more than one survey will be incorporated into a seismic interpretation of an area. One of the interpreter's first tasks in such a situation is to prepare base maps at a suitable scale for the interpretation in hand. A broad reconnaissance study might be mapped at 1:250,000 scale, a regional mapping exercise at 1:100,000, a more detailed mapping investigation where a dense seismic grid and drilling data are available at 1:50,000, and a very detailed field study at 1:10,000. In some areas, field survey maps at non-metric scales may be used during data acquisiton, but these are usually adjusted to an appropriate metric scale for interpretation work.

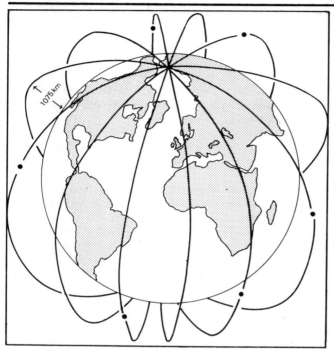

Figure 2/33: Orbit diagram for the Transit navigation satellites (after McQuillin and Ardus, 1977).

by reference to geodetic surveying of the area and a datum point which is assigned an appropriate latitude and longitude value. In Europe, the European Datum is used, and the actual location of the datum point is the Helmert Tower near Potsdam. The North American datum block which extends from Panama to the Canadian Arctic is referred to a fixed point at a location called Meades Ranch. Other areas of the globe are mapped with reference to other internationally recognised datum points. Thus any map should state not only the projection but also the datum to which topographic positions are related. A further complication is that as mapping systems have developed, different mapping agencies have adopted different spheroids as a means of computing the relationship between co-ordinate system, topographic position and latitude and longitude. The translation of positional data, mapped using one system into another can therefore be a complex problem, though easily handled by modern computers. The interpreter's principal concern is that of checking data to ensure that there is complete consistency in the mapping methods used, and that where adjustments have been necessary, these are accounted for and documented. As an example of the magnitude of errors which can arise if proper adjustments are not made, in the North Sea datum shifts between the British National Grid system, UTM using the European Datum and Satellite data are of the order 50—100m.

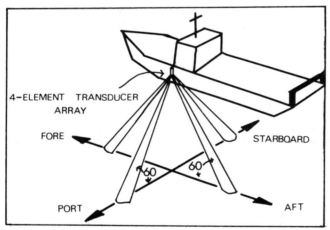

Figure 2/34: A 4-element doppler sonar (after McQuillin and Ardus, 1977).

Scale adjustment may not be the only problem. It is possible that different surveys within an area, or surveys adjacent to each other have been mapped using different map projections, or co-ordinate systems and based on different central meridians, datums and spheroids. Therefore information on map projection etc should be checked before mapping data from different surveys.

The most common map projection used in offshore exploration is the Transverse Mercator projection, and there is now wide acceptance of the Universal Transverse Mercator (UTM) system in exploration work. In such a system a relationship is defined mathematically between a grid co-ordinate system and latitudes and longitudes. Position with reference to local topography is then specified

References and suggested reading

F.T. Allen, 'Some characteristics of marine sparker seismic data' *Geophysics,* 37, (1972) pp. 462-70.

J.W. Bedenbender, R.C. Johnston and E.B. Neitzel, 'Electroacoustic characteristics of marine seismic streamers' *Geophysics,* 35, (1970) pp. 1054-72.

H.H. Bybee, 'Navigation satellites for geophysical exploration' *Offshore Technology Conference,* paper 1785 (1973).

J.R. Cole, 'Vibroseis — effective, harmless exploration tool'. *Oil and Gas Journal,* 30 October 1976 pp. 97-108.

A.T. Dennison, 'The design of electromagnetic geophones. *Geophys. Prosp.'* Vol. 1, (1953) pp. 3-28.

B.S. Evenden, D.R. Stone and N.A. Anstey, *Seismic prospecting instruments.* (Gebruder Borntraeger, Berlin).
Vol. 1: *Signal characteristics and instrument specifications* by N.A. Anstey (1970).
Vol. 2: *Instrument performance and testing* by B.S. Evenden and D.R. Stone (1971).
These volumes give a full account of seismic instrumentation design and performance.

B.E. Giles, 'Pneumatic acoustic energy source' *Geophys. Prosp.,* 16 (1968), pp. 21-53.

F.S. Grant and G.F. West *Interpretation theory in applied geophysics.* (McGraw-Hill Book Company, New York, 1965).
A mathematical text on interpretation methods used in applied geophysics.

J.T. Hornabrook, 'Seismic re-interpretation clarifies North Sea structure.' *Petroleum International*, April-May, 1974.

F.S. Kramer, R.A. Peterson and W.C. Walter, (eds). *Seismic energy sources — 1968 Handbook.* (Bendix United Geophysical, Pasadena, 1968).

R. McQuillin and D.A. Ardus, *Exploring the geology of shelf seas.* (Graham and Trotman Ltd., London, 1977).

W.H. Mayne, 'Common reflection point horizontal data stacking techniques.' *Geophysics,* 27 (1962), pp. 927-38.

A.W. Musgrave, *Seismic refraction prospecting.* (Society of Exploration Geophysicists, USA, 1967). An edited collection of papers covering the theory, applications and interpretation of seismic refraction prospecting.

P. Newman 'Water gun fills marine seismic gap.' *Oil and Gas Journal,* 7 August 1978.

P. Newman, P. Haskey, J.O. Small and J.D Waites, 'Theory application of water gun arrays in marine seismic exploration' Paper presented to 47th meeting of SEG, Calgary, 1977.

M. Schoenberger, 'Optimisation and implementation of marine seismic arrays' *Geophysics*, 35, (1970) pp. 1038-53.

Seismos Prakla, *Data acquisition — offshore geophysics,* (Hanover).

L. Seims and F.W. Hefer, 'A discussion on seismic binary-gain-switching amplifiers' *Geophys. Prosp.*, 15 (1962) pp. 23-34.

R.E. Sheriff *Encyclopaedic dictionary of exploration geophysics.* (Society of Exploration Geophysicists, USA, 1973). A well illustrated encyclopaedia covering the whole field of exploration geophysics.

W.M. Telford, L.P. Geldart, R.E. Sheriff and D.A. Keys, *Applied geophysics.* (Cambridge University Press, Cambridge, 1976). A comprehensive text-book of applied geophysics providing a valuable work of reference for exploration geophysicists.

3. DATA PROCESSING

In this chapter we shall describe a typical sequence for the processing of multi-channel seismic data. Each individual processing contractor has its own particular packages of computer programs for this purpose; we shall not consider the details of any particular system, which are in any case subject to continuous improvement, but rather discuss the basic principles involved. The object of the processing is to convert the raw data on magnetic tape, obtained as described in chapter 2, into a seismic section showing the disposition of seismic reflectors along the transverse. We describe the various stages of processing in the order that they might be carried out in practice; the main description is written in terms of marine data but the differences met with in land data are dealt with as they arise. A block diagram is shown in figure 3/1.

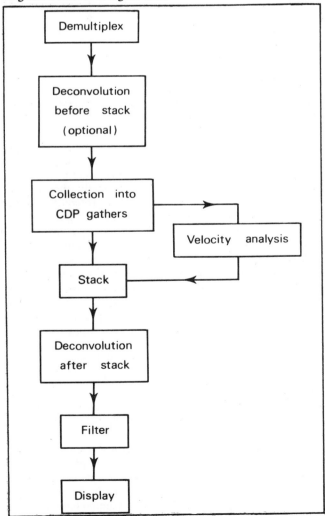

Figure 3/1 Generalised processing sequence of multi-channel reflection seismic data.

3.1 Demultiplexing and amplitude manipulation

As explained in chapter 2, the raw data are recorded in multiplexed form; the first step is therefore to re-order the data so that the series of samples corresponding to each separate geophone are brought together. After this 'demultiplexing' stage we have on magnetic tape a separate trace for each geophone at each shot point, sampled at whatever interval has been used during recording (usually 4msec).

The next step is to correct for the different gain used by the recording equipment at different reflection times; as explained in chapter 2, the amplifier gain for each sample is recorded on the original field tape, and this information is used to reproduce the amplitudes that would have been observed at a constant gain setting.

Next it is necessary to correct for spherical divergence. The source produces a spherical wave, and as the wave front spreads out the amplitude decays because the available energy is spread more thinly over an expanded wave front. In a medium of constant velocity, the amplitude is proportional to $1/z$, where z is the distance from the source. In practice, velocity usually increases with depth, leading to an additional spreading of the ray-paths near normal incidence, and a faster decline of amplitude with distance (see figure 3/2). To correct for this effect, we need to know the relation between velocity and depth; at this stage of the processing we have no detailed information and it is usual to employ an estimated time-depth relation possibly obtained from previous work in the area. If no other information is available the data may be scaled by the square of the travel time.

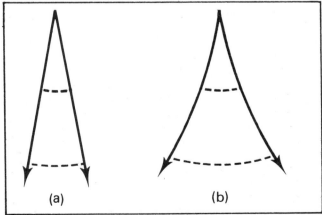

Figure 3/2: Spherical divergence: **a)** constant velocity medium, **b)** velocity increasing with depth.

Finally, each trace is scaled so that all traces have the same average amplitude. This corrects for minor variations in source amplitude and geophone sensitivity.

Before further processing, it is also necessary to remove the first arrivals from the traces; for the farther traces, these generally are due to refracted waves rather than reflections. We shall see later that shallow events recorded on the far geophones would not in any case stack correctly (see p.47). The effect of blanking off or 'muting' these first arrivals is that the first event seen on a marine seismic section is sometimes well below the theoretical time to sea-bed, especially on older data. Occasionally, careless muting is a cause of apparent bands of noise in the shallow part of the section.

3.2 Static or datum corrections
Before using seismic times either direct from field records, or more likely seismic sections, these times must be adjusted to an arbitrary datum to allow for anomalous effects induced by significant elevation and/or near surface velocity changes. Where static corrections vary widely, the quality of the final stack and the error margin of line-ties will depend on their accuracy.

3.2.1 Marine static corrections
In marine work, with both the source and hydrophone cables being very close to a constant datum, sea-level, it seems obvious to apply corrections, but it is surprising how often it is not done (and also not so declared, in the section header). When applied the correction consists of adding the source and cable depths (below the theoretical mean sea-level: not the observed sea-level which may be taken in high seas at high tide-levels) and dividing by the sea-water velocity. As a rule of thumb many geophysicists use 5000ft/s or 1524m/s, but water velocity varies widely with temperature and salinity and this quoted value would apply to temperatures in the region of 70°F (21°C) at normal sea-salinities of 35 parts per thousand. The actual velocity equation is:

$$V = 1449 + 4 \cdot 6T - 0 \cdot 05T^2 + 0 \cdot 0003T^3$$
$$+ (1 \cdot 39 - 0 \cdot 012T)\ (S - 35) + 0 \cdot 017Z$$

where V is in m/s, T is °C and Z is metres below the sea surface.

Corrections for sea-floor topography are generally not called for; any steep dip effects or buried channel anomalies can be noted and adjusted for at the interpretation stage.

3.2.2 Land statics: shots located in bedrock
Most land seismic surveys will encounter variable surface deposits ranging from clays through sands and gravel to glacial boulder clays, ice and permafrost. If dynamite is used, where it is economically practicable, shot-holes should be drilled into solid bedrock and charges set there. Shots below weathering zone can be simply corrected to a datum plane (see figure 3/3). Corrections consist of elevation and uphole computations. The time correction T_D to datum E_D is given by:

$$T_D = \left(\frac{2E_D - (E_S - D_S) - (E_G - D_{SG})}{V_B} \right) - T_{GUH}$$

where E_S is the true elevation of the ground surface at shot-hole location,
$\quad D_S$ is the depth of shot beneath ground surface.
$\quad E_G$ is the elevation of the geophone station,

D_{SG} is depth of a shot-hole located near the geophone station, and

T_{GUH} is the uphole time measured at the geophone location: this is measured by placing a shot at depth D_{SG} and a geophone at the surface at location E_G and measuring the time between initiation of shot and the time-break measured at the geophone. If no shot-hole is sited at a particular geophone station, an uphole time T_{GUH} can be interpolated from uphole time measurements at adjacent shot-holes.

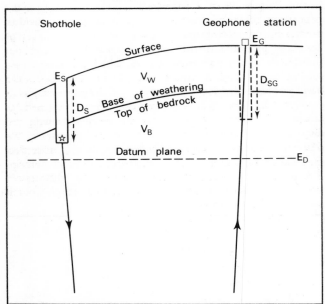

Figure 3/3: Elevation and weathered layer corrections in relation to shotholes and uphole time measurements.

It should be noted that in the equation, E_D is the height difference between the local datum plane being used for adjustment of statics and the regional datum to which E_S, E_G and E_D are referenced. Generally the regional datum is taken as sea-level and sea-level referenced elevations will be used. The velocity of bedrock, V_B is obtained from a deep uphole survey which is always conducted in each new area. A deep hole is drilled to a point substantially below the base of the interface between weathered layer and bedrock and shots are fired at various levels in the hole to provide a plot of depth against time. The slope of the curve associated with bedrock shots gives the required velocity. An alternative method, where borehole caving might be a problem, is to shoot a series of shots at ground level with a hydrophone located at various levels in the borehole. A field example of an uphole survey plot can be seen in the upper right-hand inset of figure 3/4.

3.2.3 Land statics: shots located in weathered zones
Where it is impracticable to locate all shots in bedrock, due either to the thickness of the weathered zone or the use of non-explosive seismic sources, refraction theory must be utilised. First break times of the records are plotted against distance as shown in the actual field example of figure 3/4. In the simple schematic case, illustrated in figure 3/5, the slope $1/V_W$ is the reciprocal of the weathered layer velocity V_W and the slope $1/V_B$ is the reciprocal of the bedrock velocity V_B. Where the two lines intersect at the cross-

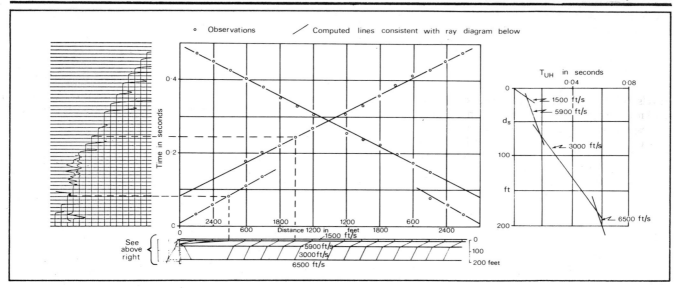

Figure 3/4: Field example of the use of refraction theory to correct for weathered layer (adapted from K.E. Bung, 'Exploration problems of the Williston Basin' *Geophysics* 17 (1952) No.3).

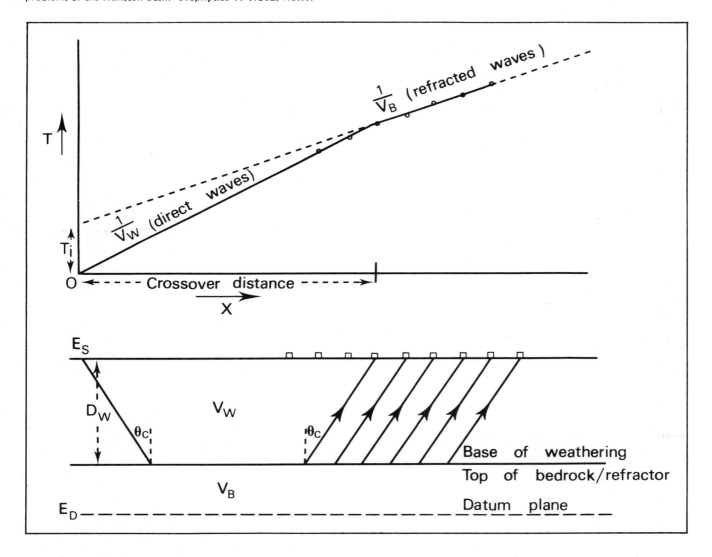

Figure 3/5: Schematic layout of refraction survey to determine weathered layer thickness and velocity.

over distance, there is onset of first arrivals from bedrock. The refracted waves line is projected back to the zero X-axis to obtain in intercept time T_i. If the horizontal source-receiver distance is X, then clearly the travel time T is given by:

$$T = \frac{X - 2D_W \tan \theta_C}{V_B} + \frac{2D_W}{V_W \cos \theta_C}$$

Thus at $X = 0$,

$$T_i = \frac{2D_W}{\cos \theta_C}\left(\frac{1}{V_W} - \frac{\sin \theta_C}{V_B}\right) \text{ and since by Snell's Law,}$$

$$\theta_C = \sin^{-1}\frac{V_W}{V_B},$$

$$T_i = \frac{2D_W \cos \theta_C}{V_W},$$

thus, by substitution:

$$T_i = \frac{2D_W\sqrt{V_B{}^2 - V_W{}^2}}{V_B V_W} \qquad (2)$$

By plotting all first breaks, and inserting the appropriate values of T_i, V_B, and V_W, depth to bedrock refractor can be computed and a datum correction applied:

$$T_D = \frac{-2D_W}{V_W} + \frac{2(E_D - E_S + D_W)}{V_B} \qquad (3)$$

where E_S and E_D are the surface and datum plane elevations. Here it is assumed that both shot and geophone are at ground surface. The T against X plot will vary for each shot location and spread as values of V_W, V_B, T_i, D_W and T_D vary. In multifold shooting, there are generally enough shot-points within each spread to allow interpolation of T_D values giving adequate definition of the weathered layer along the entire length of spread. Where elevation changes are significant, an additional correction can be applied correcting for elevation variation, in this case:

$$T_E = \frac{E_G - E_S}{V_B}. \qquad (4)$$

If shots are located in boreholes, an uphole time T_{UH} will have been measured at each shot location and this should be incorporated into the correction; it will be subtracted from the total correction time, and correspondingly the shot depth D_S is subtracted from D_W.

The above refraction corrections all assume idealised and simple earth structure. Severe elevation changes can be particularly troublesome, affecting the refractor plot and therefore the intercept time: it may be necessary to make the elevation and uphole corrections before constructing plots of T against X, especially where good data exist on weathered layer velocities.

3.2.4 Plus-minus refraction corrections

Among other available refraction methods, a particularly elegant example is the plus-minus method of Hagedoorn*. It is particularly suited to the volumes of first-break data provided by modern multifold acquisition methods, and can be easily adapted to computer processing.

* Hagedoorn 1959. 'The plus-minus method of interpreting seismic refraction sections'. *Geophysical Prospecting*, 7(1959), pp.158-182.

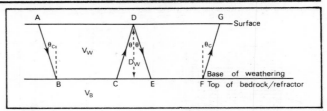

Figure 3/6: Principle of the plus-minus method of weathered layer correction.

Its principle is simply illustrated in figure 3/6. If A and D are source-points and refracted breaks are measurable at D and G, 'plus' and 'minus' values are given by the following relationships of refraction times:

$$\text{Plus time} = T_{ABCD} + T_{DEFG} - T_{ABFG}$$
$$\text{Minus time} = T_{ABCD} - T_{DEFG} - T_{ABFG}$$

By considering all the ray-path segments, it can be easily shown that the plus time = intercept time.

As defined earlier, $\quad T_i = \dfrac{2D_W \cos \theta_C}{V_W}$

and therefore $\quad D_W = \dfrac{T_i V_W}{2 \cos \theta_C}$

$$= \frac{(\text{Plus Time})V_W}{2 \cos \theta_C} \qquad (6)$$

or $\qquad = \dfrac{(\text{Plus Time})V_W}{2\sqrt{1 - \left(\dfrac{V_W}{V_B}\right)^2}}$

Similarly it can be established that the 'minus' values are related to the reciprocal velocity straight lines of the T/X plots. Plotting of the values will therefore establish the refraction velocities. The datum corrections are then computed by the same procedure as indicated previously in equations (2) through (4). Note again that if elevation changes are severe, uphole and elevation corrections may be required before plotting and computation.

The plus-minus method can be employed in cases where (as is often the case) the base of the weathering is an undulating surface, in which case the simple travel-time plots are likely to be hard to interpret.

3.2.5 Automated static picking

As mentioned earlier, any of the datum correction methods produce approximate solutions which are far from unique. It is therefore necessary to attempt to adjust for residual static problems during processing at the playback computer centre.

One method employed is to cross-correlate the best trace of a CDP gather with the other traces of the gather, thereby determining a time shift necessary for each trace to be in alignment with the others in preparation for final stack. It will be readily apparent from figure 3/7 that consistent static shifts should be expected for all ray paths with common source point, and all those with a common receiver. Static shifts will also contain increments for

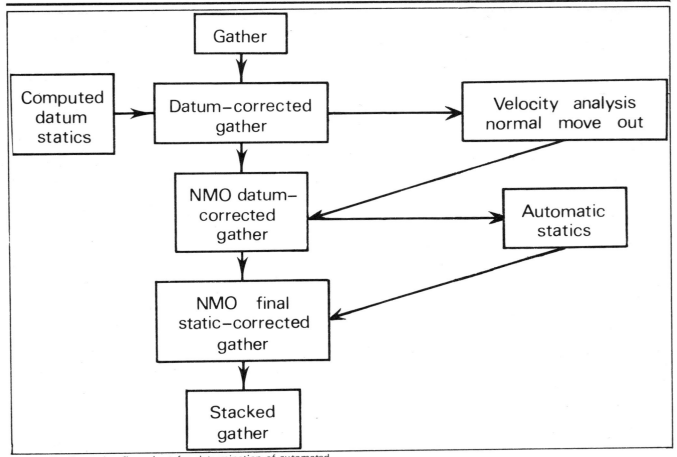

Figure 3/8: Processing flow chart for determination of automated statics.

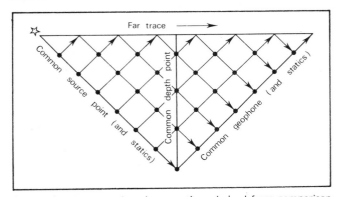

Figure 3/7: Automated static corrections derived from comparison of CDP gathers.

residual NMO (see p.43), and reflection structure or dip. These increments will be separately recognisable by their progression with distance within the gather (see p.43) for the former, and between gathers for the latter. Various different algorithms are used by contractors and oil companies. Apart from these differences the only variation between methods is in whether the datum statics are applied before or after normal moveout. Generally speaking it is preferable to do it before. The flowchart of figure 3/8 shows the sequence of processing.

3.3 Deconvolution before stack

If the source signal were an ideally sharp pulse, a series of sharp reflected pulses would be received from the various seismic discontinuities within the earth. However, we have seen in chapter 2 that such an ideal source does not exist, and the result is that a complex wave form will be reflected by each discontinuity (figure 3/9). This will tend to obscure the record as parts of wave trains from different reflections overlap.

We can regard the output wave form (ie the reflected signal actually received) as the convolution of the input wave form with the impulse response of the earth (see appendix I). To restore the output to the ideal form of figure 3/9a), we have to undo this convolution; hence the term 'deconvolution' for this process. It can be shown that, for the source wave forms of seismic interest, it is possible to design a filter which, when it operates on the source wave form, will transform it to a single pulse, although as we shall see this may not necessarily be at the time-origin of the initial signal.

Such a filter is equivalent to a convolution operator in the time domain. In the seismic case, we are dealing with discrete samples of the continuous signal, so that the following discussion is in terms of summation of a series of terms rather than of integration.

Our object is to design an operator D which when convolved with the source signal S produces a single spike δ

$D * S = \delta$ (using $*$ to denote the convolution operation).

The observed seismic reflection signal R is the convolution of the earth's impulse response E (which is the set of reflec-

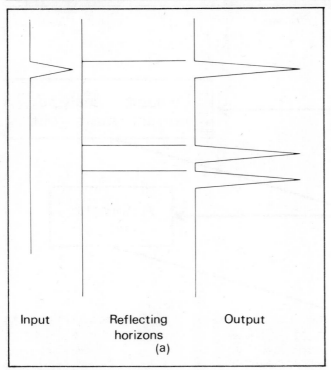

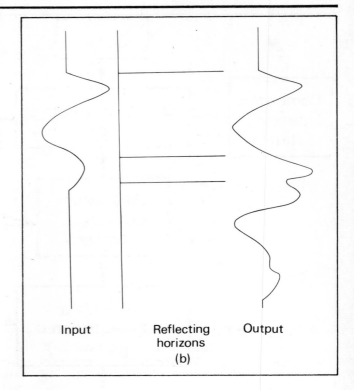

Figure 3/9: Reflection of seismic signals. **a)** input a sharp spike, **b)** input a complex waveform.

tions we actually want) with the source signal:

$$R = E * S$$

Therefore, if we apply the operator D to the observed signal, we obtain:

$$D * R = D * E * S = E * \delta = E,$$

since the convolution operation is commutative and associative, and convolution with a spike simply reproduces the original function. Therefore, if we can design a suitable operator D to collapse the source signal to a single spike, we can apply it to each observed reflected signal so as to remove the effects of source signal shape and leave only the earth's response. The first problem is how to design the operator D.

If this convolution operator is of finite length, it will not in general achieve complete pulse compression. However, it can be shown that the 'best' operator, in the sense of producing an output with the smallest least squares deviation from that desired, is given by the set of equations:

$$\varphi_{xz}(j) = \sum_{\tau=0}^{n} f_\tau \cdot \varphi_{xx}(j-\tau) \qquad j = 0, 1 \ldots n.$$

where φ_{xx} is the auto-correlation of the input,
$\qquad f_\tau$ are the convolution operator coefficients
and $\qquad \varphi_{xz}$ is the cross-correlation of the input and the desired output.

Thus, if we wish to create an operator of length 2 ($n = 1$) whose effect is to produce, as nearly as possible, a spike at the origin, we would have the equations:

$$\varphi_{xx}(0).f_0 + \varphi_{xx}(1).f_1 = \varphi_{xz}(0) \qquad (j=0, \tau=0, 1)$$
$$\varphi_{xx}(1).f_0 + \varphi_{xx}(0).f_1 = \varphi_{xz}(1) = 0 \qquad (j=1, \tau=0, 1)$$

(Note that the auto-correlation is an even function, so $\varphi_{xx}(-1) = \varphi_{xx}(1)$).

To take a concrete example, consider the wave form $(7, -3, 1)$ which might represent a decaying bubble-pulse oscillation (see figure 3/10). The auto-correlation of this input is:

$$7 \times 7 + (-3) \times (-3) + 1 \times 1 = 59 = \varphi_{xx}(0)$$
$$7 \times (-3) + (-3) \times 1 = -24 = \varphi_{xx}(1)$$
$$7 \times 1 = 7 = \varphi_{xx}(2)$$

Thus if we want the best 'spiking' operator of length 2 we find:

$$59 f_0 - 24 f_1 = 7$$
and $\qquad -24 f_0 + 59 f_1 = 0$
whence $\qquad f_0 = 0.142$ and $f_1 = 0.058$

so that the required operator is $(0.142, 0.058)$ which produces an output wave form as follows:

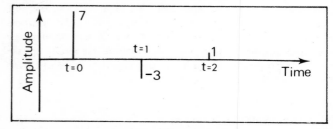

Figure 3/10: Digital sample of a simple input waveform.

$t = 0$, amplitude $= (7) \times 0.142$ $\qquad = 0.99$
$t = 1 \qquad = (-3) \times 0.142 + (7) \times 0.058 \qquad = 0.02$
$t = 2 \qquad = (1) \times 0.142 + (-3) \times 0.058 \qquad = 0.03$
$t = 3 \qquad = (1) \times 0.058 \qquad = 0.06$

Most of the amplitude of the output wave form is concentrated in a spike at the origin. Note, however, that if the waveform had been (1, -3, 7) its auto-correlation function would have been the same, whereas the cross-correlation with a unit spike at $t = 0$ would have been (1, 0, 0), leading to an operator (0.020, 0.008) which on application would give an output waveform (0.02, -0.05, 0.12, 0.06); thus spiking would not have been achieved. If we were able to take a longer operator, for example, three elements instead of two, it would be possible to achieve an even greater concentration of amplitudes in a single spike. In this case, the equations for the elements of the filter would be:

$$59 f_0 - 24 f_1 + 7 f_2 = 7$$
$$-24 f_0 + 59 f_1 - 24 f_2 = 0$$
$$7 f_0 - 24 f_1 + 59 f_2 = 0$$

which can be solved to yield the operator (0.145, 0.062, 0.008). The effect of this operator on our wave form is to produce the result (1.015, -0.001, 0.015, 0.038, 0.008) where the concentration of energy in the initial spike has been improved somewhat over the two-element operator.

Thus, if we knew the form of the signal input, we could design a suitable operator to apply to the seismic record, whose effect would be to collapse each reflection into a good approximation to a single spike. With some seismic systems, such as 'Maxipulse' it is usual to record the input signal for each shot. However, the 'input' signal varies somewhat in its passage through the earth due to selective absorption of high frequencies, and it is desirable to extract the required information from the seismic trace itself. All that is required to be known is the auto-correlation function ϕ_{xx} of the input wavelet; it can be shown that provided that the reflection coefficient within the earth is a random sequence and that the signal to noise ratio is high enough for noise to be neglected, this is equal to the auto-correlation of the seismic trace. In practice, selected seismic traces are examined in turn; for each section of the trace (design gate) the auto-correlation function is calculated and a suitable deconvolution operator deduced which is then applied to that section of the record. At this stage of processing, the operators are typically quite short, perhaps 100msec, because the seismic input signal will generally be a fairly sharp pulse. In the case of dynamite sources like Maxipulse, where bubble oscillations may persist through several cycles, it is however necessary to use much longer operators, perhaps approaching one second in length, to achieve satisfactory spiking, and these operators may have to be applied on a trace-by-trace basis before any amplitude manipulation has been effected.

There is a tendency to degrade the signal to noise ratio of the data as the spiking becomes more effective; this is because the deconvolution operation attempts to produce an output with the same amplitude in each frequency band, including those where no source signal is present. A balance has therefore to be struck between spiking effectiveness and noise introduction. It is possible to design filters which limit deconvolution to those frequencies which contain useable signals. It is also advantageous to retain some character in the seismic events rather than try to compress them all into sharp spikes. For this reason it is usual to design an operator which leaves the auto-correlation function unchanged up to its first or second zero crossing.

3.4 Collection into CDP gathers

At this stage, the traces are still ordered so that for each shot point, all 24 or 48 channels are recorded in sequence. The traces are now re-ordered so as to bring together all traces pertaining to one common depth point; this collection of traces is called a CDP gather (see figure 3/11).

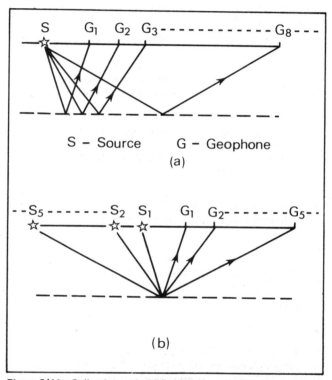

Figure 3/11: Collection of CDP gathers: a) data as recorded, b) as reordered to form a common depth point gather (CDP gather).

3.5 Velocity analysis

Before proceeding further, it is necessary to analyse the data to obtain information about seismic velocities along the section. In principle, each gather contains velocity information, but in practice it is usual to select positions for velocity analyses every few kilometres along the section. If possible, these analysis sites should be chosen having regard to the geology of the area; locally anomalous regions such as fault planes should be avoided.

The essence of the method is that the more oblique ray-paths of the CDP gather (figure 3/11b) will have longer travel-times than those near normal incidence, and the time delay will depend on the seismic velocity. Consider the geometry of figure 3/12. Clearly the total path length P is given by

$$P = 2\left(\left(\frac{x}{2}\right)^2 + d^2\right)^{\frac{1}{2}}$$

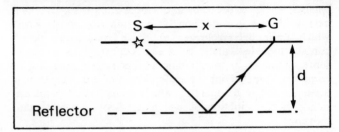

Figure 3/12: Geometry of a CDP gather in a homogeneous medium.

so if the seismic velocity is v (asumed uniform) and the two-way travel time is T,

$$T = \left(T_o{}^2 + \frac{x^2}{v^2}\right)^{\frac{1}{2}} \text{ (a hyperbolic relationship)}$$

where $T_o = \dfrac{2d}{v}$ is the vertical incidence two-way time.
If $xv \ll T_o$, then
$$T \doteq T_o + x^2/2T_o v^2 \text{ (the parabolic approximation)}.$$

The reflection time is thus increased by an amount $x^2/2T_o v^2$, which is called the normal move-out (NMO). The size of this term decreases down the seismic record, for example for

$d = 2km, v = 3km/sec, T_o = 1\frac{1}{3}\text{ sec, so for } x = 2km,$
NMO $\doteq$ 170msec,
$d = 4km, v = 3km/sec, T_o = 2\frac{2}{3}\text{ sec, so for } x = 2km,$
NMO $\doteq$ 80msec.

It is worth noting that in the CDP gather, the effect on the NMO of dip on the reflector is rather small (figure 3/13). Here the total path length is given by:

$$P^2 = S'G^2 = SS'^2 + x^2 - 2x \cdot SS' \cdot \sin \alpha.$$
But $\quad SS' = 2(d + x/2 \sin \alpha)$,
therefore $P^2 = x^2 \cos^2 \alpha + 4d^2$.
Thus $\quad T^2 = T_o{}^2 + \dfrac{x^2 \cos^2 \alpha}{v^2}$

with the same notation as before, T_o being the two-way time perpendicular to the reflector. The effect of the dip is therefore to increase the apparent velocity of a factor see α, which is close to 1 for small dips.

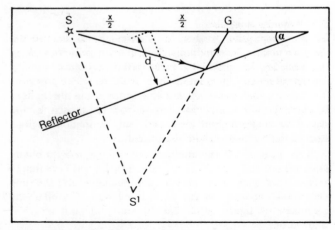

Figure 3/13: Geometry of CDP gather for a dipping reflector. S^1 is the mirror image of S in the reflector.

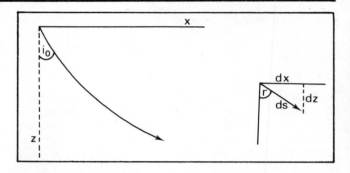

Figure 3/14: Refraction effect of downward velocity increase on seismic ray.

If the seismic velocity is not constant with depth, the analysis proceeds as follows (figure 3/14). According to Snell's Law,

$$\frac{\sin i}{\sin r} = \frac{v_1}{v_2}, \quad \text{(see chapter I)}$$

so that at any point the angle that the ray makes to the vertical is given by

$$\sin r = \frac{v}{v_o} \cdot \sin i_o$$

where i_o is the initial angle of incidence and v_o is the initial velocity (ie at z = 0).

Writing $\quad u = \dfrac{v_o}{\sin i_o}, \quad \sin r = \dfrac{v}{u}$

The path length of a segment of ray is given by

$$ds = \frac{dz}{\cos r} = \frac{dz}{\sqrt{1 - v^2/u^2}}$$

The time taken to transverse this segment is

$$dt = \frac{dz}{v\sqrt{1 - v^2/u^2}}$$

The total travel time is therefore

$$T = \int \frac{dz}{v\sqrt{1 - v^2/u^2}}$$

and $T_o = \int \dfrac{dz}{v}$ because $u \to \infty$ for vertical incidence.

The x distance travelled on the segment is given by

$$dx = dz \tan r = \frac{dz \cdot v/u}{\sqrt{1 - v^2/u^2}}$$

Therefore, the total x distance is given by

$$X = \int \frac{v \cdot dz}{\sqrt{u^2 - v^2}}$$

To calculate the NMO we proceed as follows in the parabolic approximation; v/u will be small near normal incidence so

$$T = \int \frac{dz}{v} \left(1 + \tfrac{1}{2} \cdot \frac{v^2}{u^2} \cdots \cdots \right)$$

$$\doteq T_o + \tfrac{1}{2} \int \frac{v \cdot dz}{u^2}$$

But consider

$$S = \frac{x^2}{2 \int v^2 \, dt} = \frac{\left[\int \frac{v dz}{\sqrt{u^2 - v^2}} \right]^2}{2 \int \frac{v dz}{\sqrt{1 - v^2/u^2}}}$$

$$= \frac{1}{2u^2} \int \frac{v dz}{\sqrt{1 - v^2/u^2}} \doteq \frac{1}{2u^2} \int v dz$$

Therefore

$$T \doteq T_o + S = T_o + \frac{\tfrac{1}{2} x^2}{\int v^2 \, dt}$$

This result may also be shown to be true even if the parabolic approximation is not made.

This implies that in the case where the velocity varies with depth, we can use the same expression as before for the NMO provided we interpret the velocity as being the root-mean-square velocity with respect to time. As we shall see shortly, the seismic velocity is obtained from measurement of the NMO, and therefore is the rms velocity to a particular horizon. If we wish to calculate the interval velocity between two horizons, we can proceed as follows (figure 3/15).

Clearly, $v_{rms2}^2 = \dfrac{v_{rms1}^2 \cdot T_1 + v_{interval}^2 \cdot (T_2 - T_1)}{T_2}$

or, rearranging,

$$v_{interval} = \left[\frac{T_2 \cdot v_{rms2}^2 - T_1 \cdot v_{rms1}^2}{T_2 - T_1} \right]^{\frac{1}{2}},$$

the Dix interval velocity formula.

In order to determine the rms velocity, we could simply time the arrival of a particular reflection at each geophone and plot a graph of T^2 against X^2, whose gradient would give $1/v^2$. In practice more automated methods are

necessary because of the large volumes of data involved. One technique which is sometimes used is the constant velocity gather. In this method, it is assumed that the seismic velocity has a constant specified value throughout the ray-path, and the NMO as a function of reflection time is calculated for each trace of a gather. These NMO values are used to correct the traces for the shooting geometry, and the resulting gather is displayed as a set of wiggle traces. If the true rms velocity to an event is the same as that used in the analysis, it will appear horizontal; if the true rms velocity is less than that assumed, events will bend downward toward the further traces (under-correction), and if greater, upwards. Thus we can estimate the depth where our arbitrary assumed velocity is equal to the true rms velocity. If we repeat this analysis for a series of different velocities, we can build up an rms velocity profile. The main disadvantage of this method is that on poor data alignments across unstacked traces are difficult to see. It is also possible to estimate velocities by stacking the data (see next section) assuming a constant velocity; strong events will tend to appear at those depths where the assumed velocity equals the true rms velocity, and by repeating the process for a suite of assumed velocities, a velocity profile can be prepared. In practice, this constant velocity stack can be very expensive in computer time. It has the disadvantage that, if constant velocity increments are used, the display is asymmetrical, with a rapid increase of amplitude on the low velocity side of the display up to the maximum, followed by a slow decrease on the high velocity side.

Another method is to consider separately a large number of normal incidence reflection times using perhaps a 25msec increment. At each reflection time, the NMO for each trace of the gather is calculated, using a wide spread of different rms velocities. A window (say 50msec in length) is then moved across the gather on each NMO hyperbola in turn, and the degree of correlation of the traces within that window is calculated. The results can be presented in various ways, of which figure 3/16 is an example. Here the degree of correlation has been contoured as a function of the rms velocity and two-way time. Such a display is interpreted by hand, the main peaks being joined together by straight lines; the strong peaks around 4800ft/s have been ignored as they are due to seabed reverberation multiples, whose NMO is that appropriate to the velocity of sound in water. Towards the bottom of the section it is usual to find no significant peaks owing to the lack of strong reflections; in this situation, a constant interval velocity can be employed. It generally makes little difference to the processing of the data what assumptions are made about velocities at depths below about 3—4 seconds, since the NMO is usually small at such depths.

Having made initial picks of the velocity data, it is useful to check them for consistency. A possible method is shown in figures 3/17 and 3/18. In the latter the rms velocities have been converted to interval velocities and the interval velocities have been contoured for all the velocity analyses along the line. It is then possible to see whether any analyses appear anomalous; if they do, and no reasonable geological explanation is apparent, the velocity analyses should be re-examined to find the cause and rectify it. In the display of figure 3/17, individual traces of the gather have each been adjusted by the NMO corresponding to the picked velocities. If the latter are correct,

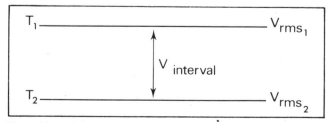

Figure 3/15: Calculation of Dix formula for interval velocity.

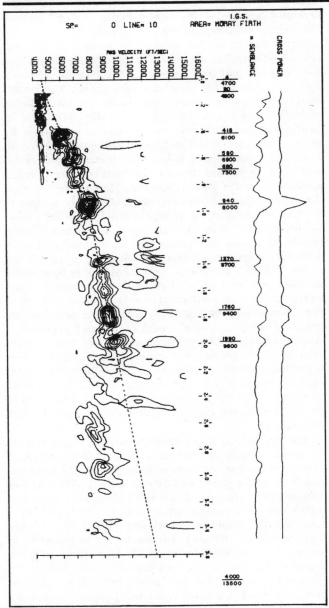

Figure 3/16: Velocity spectrum.
(*Courtesy: IGS; Seiscom survey*).

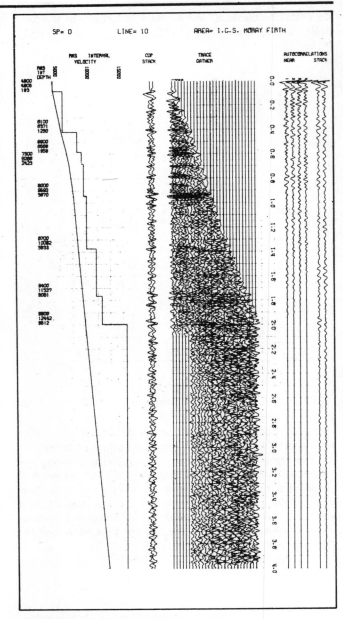

Figure 3/17: Verification of velocity analysis.
(*Courtesy: IGS; Seiscom survey*).

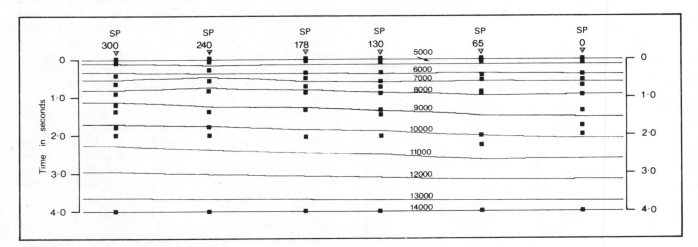

Figure 3/18: Velocity profile along a seismic line.
(*Courtesy: IGS; Seiscom survey data*).

reflections should line up horizontally across the gather. If a reflector curves downwards towards the further traces, the velocity is too high (under-correction); if upwards, too low (over-correction).

3.6 Stacking

The main point of recording data multi-fold is that by adding together all the channels a worthwhile improvement in signal to noise ratio can be achieved. If all the noise were random this improvement would be equal to $\sqrt{n}$ where n is the number of channels, that is, a factor of nearly 7 for 48-fold data. As we shall see, there is also a significant improvement in some types of non-random noise.

Before the data can be added together in this way, the NMO correction must be applied to all channels. It is for this reason that so much attention must be given to the prior calculation of velocities, for at least in the upper part of the section it is essential that this correction should be accurate.

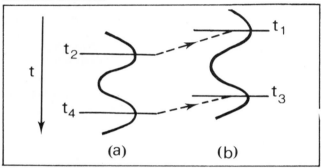

Figure 3/19: Change of frequency content during NMO correction: the two adjacent peaks a) before NMO are of higher frequency content than b) after NMO correction.

Between velocity analyses, rms velocities for each trace are calculated by interpolation. Unfortunately, the variations of NMO correction with T_0 is sufficiently great to produce appreciable alteration of frequency content for the far traces at shallow depths. The effect can be thought of as follows (figure 3/19). If two adjacent peaks of a cycle are adjusted for NMO, the deeper signal will be adjusted less, for NMO decreases with depth (see above). Therefore, the NMO-corrected signal is of lower frequency than before adjustment. This effect can be large. We can calculate its magnitude as follows:

Since $\qquad T_o = \sqrt{T^2 - x^2/v^2}$,

then $\qquad dT_o = \dfrac{T \cdot dT}{\sqrt{T^2 - x^2/v^2}}$.

But frequency $\omega = 1/t$, so $d\omega = -1/t^2 \cdot dt$ where t is the period,

therefore $\quad \dfrac{d\omega}{\omega} = \dfrac{-1}{dT} \left[\dfrac{T \cdot dT}{\sqrt{T^2 - x^2/v^2}} - dT \right]$,

since $dt = dT_o - dT$ if we imagine the sound peaks originally at time spacing dT to have been stretched to spacing dT_o by NMO correction.

Therefore $\dfrac{d\omega}{\omega} = 1 - \dfrac{\sqrt{T_o^2 + x^2/v^2}}{T_o} = 1 - \sqrt{1 + x^2/T_o^2 v^2}$

Thus there will be a 40 per cent frequency change for a two-way time of 0.5 second at a geophone distance of 1km with velocity of 2km/sec. Such large frequency changes alter the character of the signal completely and would destroy the signal enhancement of the stack. Therefore, the far traces are blanked off at shallow reflection times; this is apparent in figure 3/17.

After correction for NMO, all the traces of the CDP gather are stacked, that is, all the values corresponding to a particular reflection time on each trace are added together. Not only does this enhance the signal from true reflectors relative to random noise, but it also decreases the relative amplitude of multiples. This is apparent from figure 3/16 where strong seawater multiples are seen in the velocity spectrum with rms velocity of 4800ft/sec. Since rms velocities up to 7000ft/sec were actually used in stacking the section, these multiples are very under-corrected for NMO, and arrive later at the far geophone; therefore, they tend not to add up in phase during stacking. For example, if the maximum offset X were 1200m, at a depth T_0 of 2 seconds the NMO correction for a primary reflection of velocity 3km/sec would be 40msec, whereas a seabed to sea surface multiple would have an rms velocity of only 1.5km/sec and would need a NMO correction of 154msec to stack up. This difference of 114msec in the NMO correction implies that the multiple will tend to cancel out during stacking provided that the frequency of the signal is greater than about 1/0.114 Hz, that is, about 10Hz. On the continental shelf, this is a very effective method for the removal of multiples. In deep water however (for example in water depths equivalent to over 1 second two-way time), the rms velocities to horizons within one or two seconds of the seabed are strongly dominated by the low velocity in the water column, and the separation between rms velocities to reflectors and to multiples becomes small. In this case, stacking is not effective in multiple suppression. Stacking also fails to suppress diffraction patterns since it may be shown that the apparent NMO of a diffraction hyperbola across the CDP gather is the same (within the parabolic approximation) as the NMO to be expected of a real event.

A consequence of the muting of far traces for shallow events is that the fold of stack (that is, the number of traces added together) varies down the record. The effect of this on the amplitude of the stacked trace is corrected for by applying a time-variant scaling factor to the amplitude.

3.7 Deconvolution after stack

A further deconvolution process is usually carried out after stack. The object is to remove the reverberations caused by multiple reflections between closely-spaced strong reflectors such as the sea surface and the seabed. This is a very effective method for the removal of multiples, provided that their period is short enough for the data to provide an adequate design gate for the deconvolution operator; what precisely constitutes an 'adequate' length of data is discussed below, but in general no difficulty is experienced in removing water-bottom multiples on the continental shelf (water depths less than 200m), though there may be severe problems in oceanic water depths. Firstly it is necessary to identify the periods of such potentially damaging reverberation; these can be recognised from the auto-correlation function of a stacked trace, which has large amplitude at lag values equal to the period of major reverberation. To take an example, consider the wave form (7, −3, 1), for which we previously designed a 'spiking' operator, and suppose that it now has an attendant multiple after six time units (figure 3/20). The autocorrelation function will be (118, −48, 0, 7, −24, 59, −24, 7), which clearly shows the presence of the multiple by the high value at the lag of six units. In practice, a sectional auto-correlogram is produced, which is a composite, displayed in the form of a section, of auto-correlations of adjacent traces along a line. This will illustrate the nature and period of the multiples present.

It can be shown that the deconvolution operator of length n required to remove a multiple whose time-lag is α given by:

$$(1, 0, 0, \ldots 0, -f_0, -f_1, \ldots -f_n)$$

where there are $(\alpha - 1)$ zeros and the $f_0 \ldots f_n$ are found by the equation:

$$\varphi_{xx}(\alpha + j) = \sum_{\tau = 0}^{n} f_\tau \cdot \varphi_{xx}(j - \tau) \qquad j = 0, 1, \ldots n.$$

where ϕ_{xx} is the auto-correlation function of the input, which as with the 'spiking' deconvolution, may be taken to be the auto-correlation function of the seismic trace. Such an operator is called a prediction error filter of prediction length α, and its effect is to tend to reduce to zero the value of the auto-correlation function of the trace for all lags between α and $(\alpha + n)$. The object is therefore to choose α and n so that harmful reverberations fall between these lags.

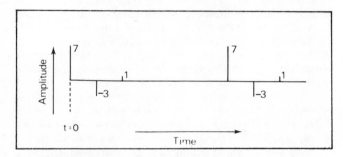

Figure 3/20: Digital waveform (of figure 3/10) with added multiple.

In the case of our example, we choose $\alpha = 6$ and $n = 1$, giving:

$$j = 0 : f_0 \varphi_0 + f_1 \varphi_1 = \varphi_6$$
$$j = 1 : f_0 \varphi_1 + f_1 \varphi_0 = \varphi_7,$$

Thus
$$118 f_0 - 48 f_1 = 59$$
$$-48 f_0 + 118 f_1 = -24$$

whence $\qquad f_0 = \tfrac{1}{2}, f_1 = 0.$

Our operator is therefore $(1, 0, 0, 0, 0, 0, -\tfrac{1}{2})$. The effect of this on the input wave form is to convert it to (7, −3, 1, 0, 0, 0, 3½, −1½, 1, 0, 0, 0, −3½, 1½, 1), thus partially removing the multiple (figure 3/21).

In practice it is usual to carry out trials on a test piece of data to determine the optimum prediction length and operator length since a balance has to be struck between reverberation suppression and the introduction of noise. Generally operator lengths of a few hundred msec are used, and the prediction distance α is set to fall short of the two-way time to the water bottom, or of the reverberation time of other harmful ringing energy. A typical rule of thumb for suppression of water-bottom multiples (in shallow water, say 100m depth) is that the operator should be approximately twice the length of the two-way time to the seabed, and should be designed using a gate approximately ten times as long as the operator length.

3.8 Filtering

As explained in chapter 1, the frequency of a reflected seismic signal tends to decrease as the path length increases owing to greater absorbtion of higher frequencies. To obtain the best signal to noise ratio, it is therefore necessary to filter the stacked section using a time-variant filter whose pass band becomes of lower frequency with increasing reflection time. Where there is appreciable structure along the profile, it is also useful to vary the time of onset of pass-band frequency alterations so as to keep the frequency content, and therefore the character, of a given event approximately constant. Frequency filters are usually chosen by inspection of narrow band-pass displays of a test section of record; from this it is possible to identify the approximate frequency content of the reflections of interest and choose filter parameters so as to include as much as possible of the frequency range of the desired events, while

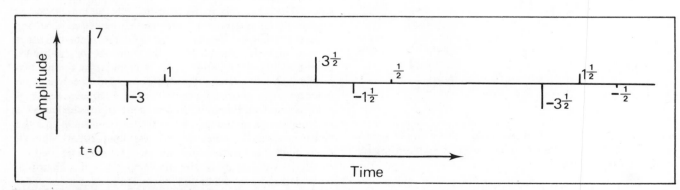

Figure 3/21: Partial removal of multiple (waveform of figure 3/20) by deconvolution.

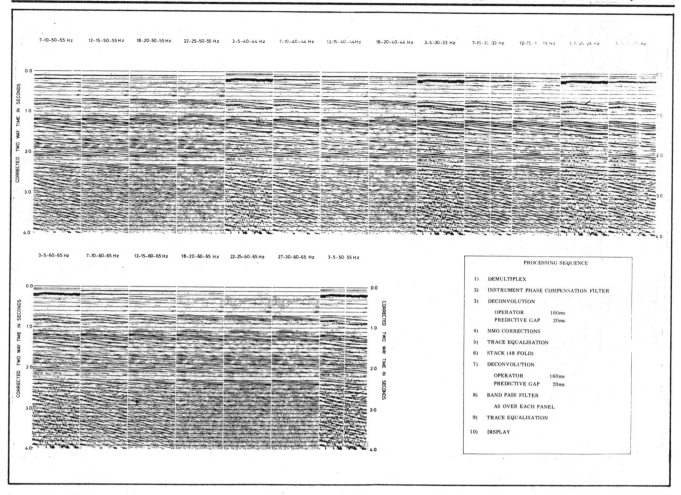

Figure 3/22: Filter test.
(*Courtesy: SSL*).

removing unwanted noise of other frequencies. An example of such a test suite is shown in figure 3/22.

3.9 Display
Before final display, amplitudes down each trace are adjusted so that that the range of amplitudes is confined to a narrow band; care is required at this stage to avoid producing dead zones adjacent to strong reflectors because of overlap of the low-gain regime into a low reflection strength region. This 'equalisation' is necessitated by the very limited dynamic range of all display modes.

Several methods of final display are currently used. They are all methods of treating a 'wiggle trace' to make the correlation between adjacent traces more visible. A very common method of display is the 'variable-area' section (figure 3/23a) in which troughs of the wiggle are blank and the peaks are shaded in solid black. Another common method is to add to this display the wiggle trace in the trough, thus giving more information about the wave form (figure 3/23b), which may be of great importance to detailed studies. It is generally possible to specify the proportion of the 'wiggle' which will be filled in so as to vary the black-white proportions to give the best effect. Figure 3/24 shows a range of display modes including variable density as well as variable area and wiggle trace presentations.

Horizontal scales are often chosen to be around 1:25,000 for the commonly used 10cm/sec and 3¾in/sec two-way time vertical scales. This produces a section which is approximately free from vertical exaggeration. For quick regional assessment, 5cm/sec scale sections are often more useful.

3.10 Migration
In areas of steep dip, reflector segments appear on the time section considerably removed from their true position (figure 3/25). This is because the reflection at CDP point X actually comes from point Y on AB, but is plotted vertically below X at Z, where $XZ = XY$ in two-way time. For steeply dipping events, there is a substantial tendency to move the segments of the event downdip (Z being downdip from Y) and to decrease the apparent dip on the section as compared with reality (since d^2 is less than d^1). A more complex situation arises when the curvature reflector exceeds that of the incident wavefront. As a result, reflections can originate simultaneously from more than one point (figure 3/26). The resulting time section is not easy to interpret in detail because although the presence of a syncline is apparent the exact shape of its sides is not at

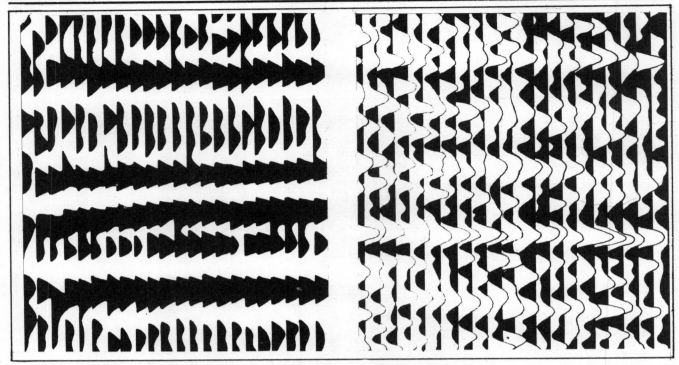

Figure 3/23: a) Variable area seismic section display, **b)** Variable area and wiggle trace section display.

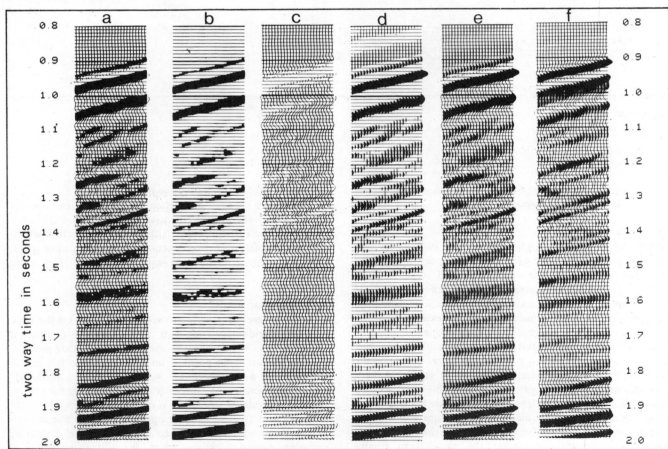

Figure 3/24: Range of section display options: Wiggle trace (c) is the normal output of a processing camera galvanometer. Variable area display (d) shades all peaks with uniform intensity. Variable density display (b) prints shading which varies with amplitude, but in this case there is no shape variation. Variable density display can be superimposed on the wiggle trace, such as in (a), similarly variable area display can be combined with the wiggle trace such as in (e). Note that (a) and (e) have had their overall intensity doubled using a two-pass system. For comparison purposes in detailed character analysis or 'bright spot' studies, a reverse polarity section (f) — here in wiggle variable area display — is often specified. (*Courtesy: GSI*).

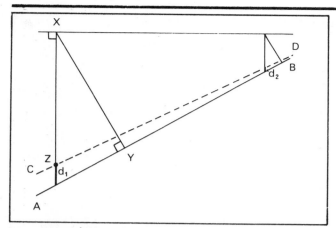

Figure 3/25: Migration: actual event AB produces an apparent event at CD.

all clear. Migration is the process by which these effects of dip and curvature of the reflector surface are removed.

We have seen in chapter 1 that a point reflector will give rise to a hyperbolic pattern on the time section, in the case where velocity is independent of depth. If velocity varies with depth, the point reflector will give rise to a curve of similar general shape, but different in detail; if the velocity-depth relationship is known, curves can be calculated for all depths of the point reflector. The reflector seen on the unmigrated section is the envelope of a suite of such curves, one for each point on the reflector surface. Thus, in figure 3/27, the true position of the point of reflection corresponding to a point A on the unmigrated section can be found by fitting one of the family of such to the slope of the reflector at A. The point of reflection is then situated at B; the slope of the reflector surface at B is given by the slope of the wavefront through that point,

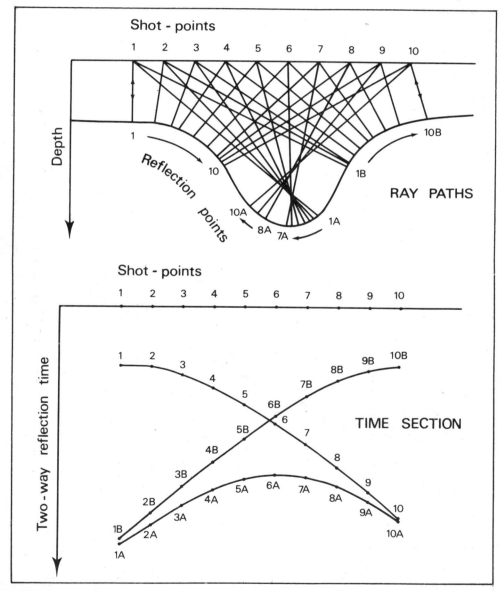

Figure 3/26: Reflections from a concave reflector; curvature of the reflector exceeds that of the wavefront. Rays emanating from each of the source locations 1-10 are reflected at up to three different points (all at normal incidence) on the concave reflector. The resulting time section shows a complex pattern of three reflector curves.

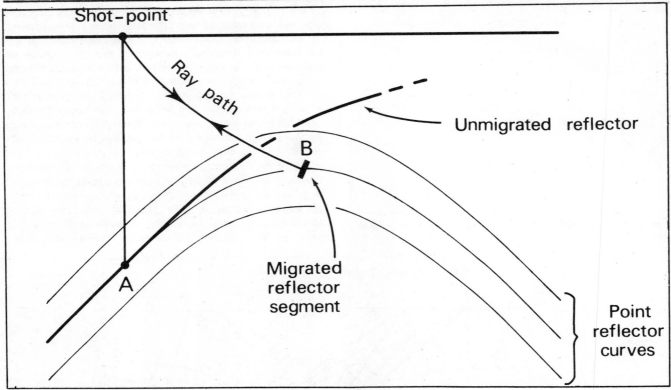

Figure 3/27: Principle of migration.

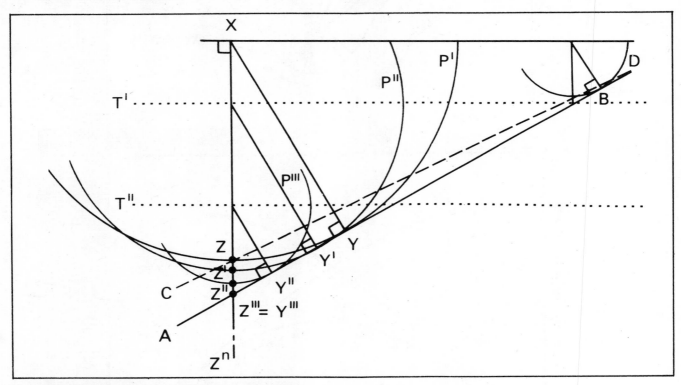

Figure 3/28: Migration: use of the reflector mapping technique.

which can be calculated from the velocity-depth relation. This process, repeated for each point of the unmigrated reflector, allows us to draw the reflector in its true, migrated configuration. It is clear that it will also collapse diffraction patterns from faults back to point reflectors. Various methods of carrying out equivalent processes by computer exist. A comparison between a migrated and an unmigrated section is shown in figure 3/30.

More recently, wave equation migration has been introduced. In this system, downward continuation of the seismic trace, recorded at the surface, is simulated by rigourous mathematical filtering algorithms; these are based on numerical solutions of the partial differential equations generated by the two-dimensional scalar wave equation, as it is acted on by a moving co-ordinate system. Essentially, movement of the source and signal receivers down through the earth is generated and wherever 'a first arrival of a downgoing wave is time coincident with an upgoing wave, reflections (are deemed) to exist at (those) points in the earth'.[*] The process is also known as 'reflector mapping'.

In figure 3/28, the apparent dipping reflector of figure 3/25, CD, is plotted, registering an apparent reflec-

---

[*] J.F. Claerbout, (1971), 'Toward a unified theory of reflector mapping'. *Geophysics*, vol.36, no.3, pp.467–481.

tion point at Z on the seismic trace of XZ^n. Until migrated, Z is a perfectly legitimate reflector point, conforming to the principle of a downgoing wave (P) coinciding with a supposed upgoing wave. Moving source and receiver to sub-surface time T', we now see a deeper apparent reflector Z', identified by the downgoing wave P' coinciding with another supposed upgoing wave: Z has now vanished. The process is repeated until source and receiver are stimulated to be at or below Z''', at which reflector point, the reality is established. Note that all points passed in the simulation play no further part in the migration. Repetition of the procedure at traces X', X'' $---$ $X^{n/}$ would similarly establish the validity of the time dipping reflector A, Z''' ($\equiv Y'''$), Y', Y, B.

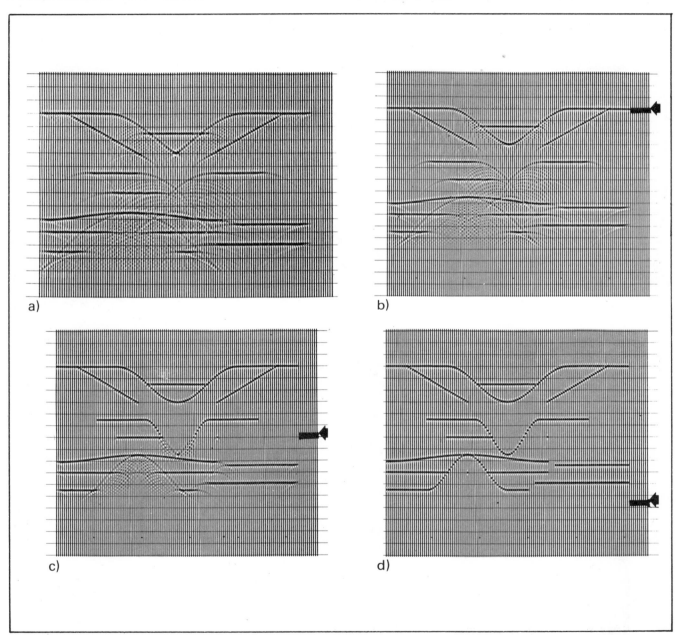

a)

b)

c)

d)

Figure 3/29: Migration: synthetic seismic models which have been wave equation migration processed. **a)** Original synthetic section, **b)**, **c)** and **d)** migrated sections with arrows indicating progressive correction down through section.
(*Courtesy: SSL*).

In the current state of the art, a moving co-ordinate system focuses the analyses on the upcoming waves only. Several different operator algorithms are offered by processing contractors, varying mainly in their attempts to reproduce the predicted original frequency and phase content.

A sequence of synthetic seismic models, wave equation migration processed, progressively downwards in time, is shown in figure 3/29. The elimination of diffraction from faults, point sources, and steep dips, and the clean accentuation of the latter is a vivid demonstration of this advanced technology.

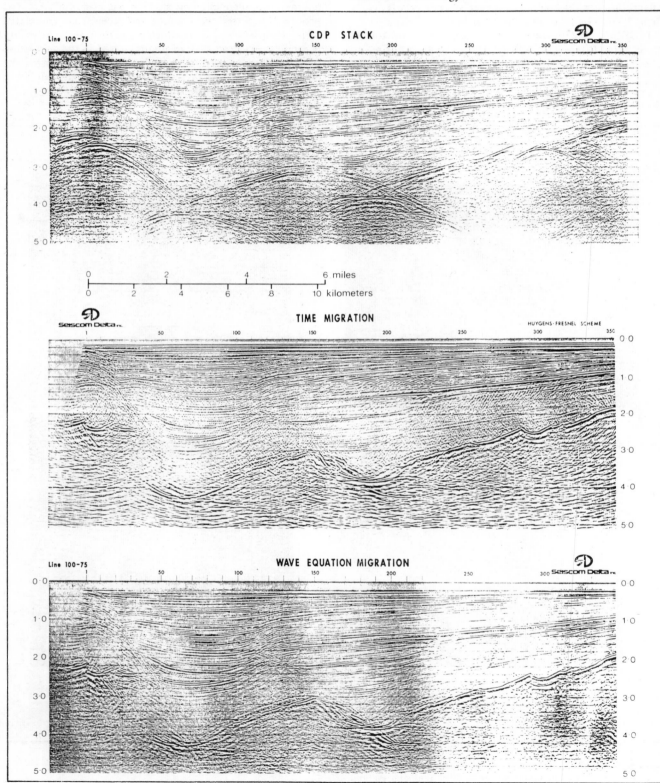

Figure 3/30: Comparison of routine stacked section with time and wave equation migrated sections from offshore Alaska. (*Courtesy: Seiscom—Delta*).

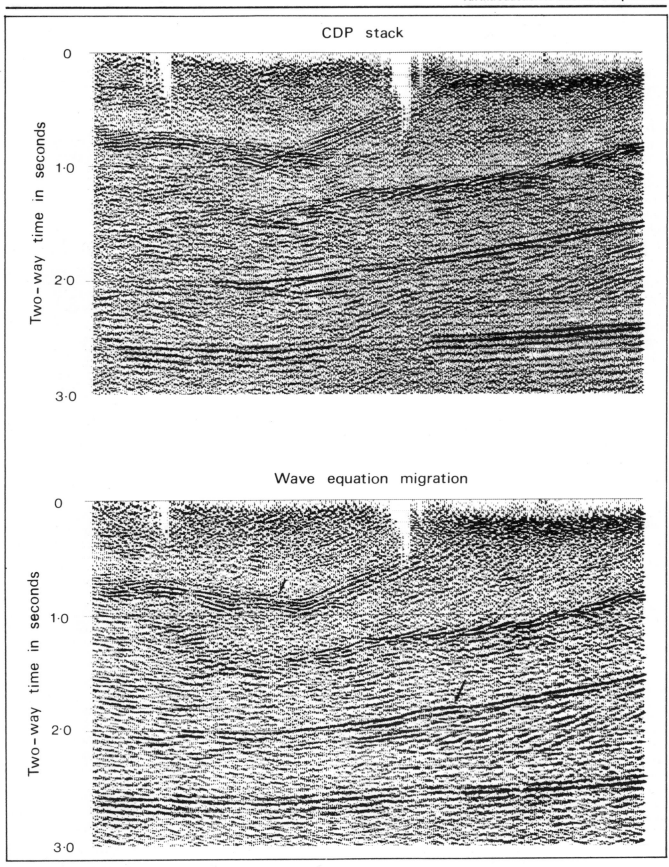

Figure 3/31: Comparison of conventional CDP stack and wave equation migrated section from the foothills area of the Canadian Rocky Mountains.
(*Courtesy: Digitech*).

The advantage of wave equation migration is that it produces less smearing than the previous method: as a consequence, amplitude and character of reflections are maintained. A field example, from offshore Alaska, figure 3/30, exemplifies this in its comparison of a wave equation migration with a time migration and with the routine stacked section. Figure 3/31 is an additional comparison of routine stack and wave equation processing. The section is from the thrust-belt of the foothills of the Canadian Mountains. Thrusted Cretaceous and Mississippian formations overlie an undisturbed Cambrian economic basement seen at approximately 2.5s two-way time. The improvement in data quality in the migrated section is clearly apparent despite the complexity of geological structure in shallow rock layers; a small fold emerges (upper arrow) within the synclinal axis and deeper in the section an upward flexure (lower arrow) of a strong reflector is both amplified and migrated to the right. Throughout, reflection continuity is much enhanced.

In general, the effect of migration is to make anticlines steeper and of smaller area, while synclines tend to be enlarged (see figure 3/33). Therefore, unmigrated sections tend to give a more optimistic assessment of the volumes of potential hydrocarbon reservoirs when located in structural traps.

3.11 Special displays

The bulk of seismic interpretation work is undertaken using conventional CDP stacked sections; even migration processing is only applied where the additional cost can be justified in terms of the degree of data enhancement achieved. In other situations however, interpretation problems can demand the application of specialised methods to, the direct detection of hydrocarbons is discussed in chapter 8.

3.11.1 Interval velocity display and depth sections

By use of the Dix equation, NMO velocities can be converted into interval velocities (see p.45). Velocity information is always useful to the interpreter, and special displays have been devised to accommodate this requirement. Overlays can be produced for each seismic section showing iso-velocity contours of interval velocity. Alternatively, colour-coded interval velocity sections (or overlays) can be produced. Figure 3/32 (p. 132) shows a conventional North Sea section over-printed with colour-coded interval velocity data. With derivation of interval velocities, it is a relatively simple matter to convert the traditional time section into one in which the vertical axis becomes a depth scale. If this process is applied to a migrated time section then the affect on structure of both time and depth migration becomes apparent. In figure 3/33 the left hand flank of an anticline is picked out by arrows at the level of a prominent reflector. This feature is seen to migrate to right between time crsss-section, migrated time cross-section and migrated depth cross-section.

3.11.2 Reverse polarity display

It is a convention on wiggle trace sections that a trough should represent a reflection at a boundary with a positive reflection coefficient. It can be extremely helpful in

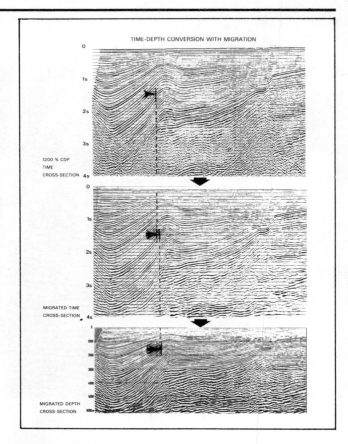

Figure 3/33: Comparison of conventional CDP stack with time migrated section and depth converted display. (*Courtesy: C.G.G.*).

detailed stratigraphic or hydrocarbon indicator studies to have seismic traces played out in reversed polarity. A comparison between normal and reversed polarity is shown in figure 3/24.

3.11.3 Seislog*

This process has been devised to operate on seismic data in such a way as to produce a pseudo-sonic log, reversing the procedure used for the derivation of synthetic seismograms. Reflection coefficients and instantaneous velocities are derived and velocites in micro-seconds per foot are displayed on a sonic log scale. Individual pseudo-sonic traces can then be interpreted using traditional log correlation techniques. In figure 3/34a a frequency analysis of a sonic log is shown. The log can be expressed as a two component system; a gross velocity framework contained in the very low frequencies upon which is superimposed a modulating series of all higher frequencies. In figure 3/34b is shown a comparison of a sonic log and a Seislog* trace. The Seislog* trace is constructed by the reversal of the frequency analysis shown in figure 3/34a. The high frequency modulating series is derived from an analysis of broad-band, conventionally processed but unfiltered and unscaled seismic data. The low frequency curve is computed from a stacking velocity function which has been low-pass filtered and smoothed. An example of the successful application of Seislog* interpretation, extending the boundaries of a gas field, is described in chapter 8.

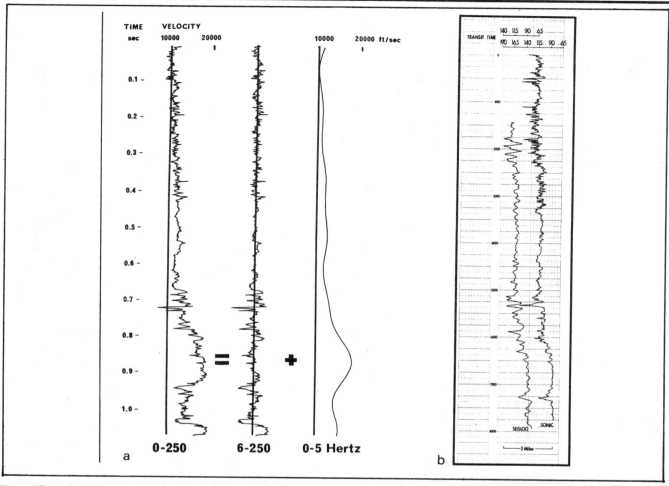

Figure 3/34: a) Frequency analysis of a sonic log. **b)** Comparison of a Seislog* and a sonic log from adjacent sites. (*Courtesy: Technika Resources Ltd*).

3.11.4 True amplitude or 'bright spot' displays

As described in chapter 2, one of the great advantages of using binary gain digital recording systems is that it is possible to record the exact gain applied to incoming signals. This implies that it should be possible to replay the actual geophone or hydrophone amplitudes. In practice however, it is desirable to maintain reflection continuity and character as well as uniform processing parameters when handling large batches of data. Thus true amplitude presentation is not readily achieved, and only applicable to specialised studies. It was the identification of significant lateral variation of reflection amplitudes in association with the occurrence of gas sands which led to the development of true amplitude or 'bright spot' presentations. The use of such displays in the detection of hydrocarbons is discussed in chapter 8. There is no single proprietry method of true amplitude processing and presentation; in the main, steps which would negate a true amplitude presentation are eliminated, and other processing parameters significantly refined. No equalisation or time variant filtering is applied. A gain recovery process is used which attempts to compensate fully for spherical divergence and absorption losses. Normal moveout velocity is determined with great care, as are the deconvolution parameters. Final sections are often produced at amplitude levels well below those of a conventional section; the effect is to give a display in which normal reflectors merge into the background,

anomalously high amplitude reflectors are high-lighted, and lateral amplitude variation at such reflectors can be clearly seen. A typical section is shown in figure 3/35. Colour can be very effective in the display of true amplitudes. For other examples of true amplitude displays see figures 8/6 and 8/7. (see pages 120 and 133).

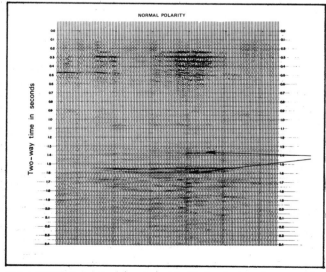

Figure 3/35: Example of a true amplitude display. Arrows point to high amplitude events which occur at both top and bottom of a hydrocarbon contact. (*Courtesy: Phoenix/SSC*).

References and suggested reading

J.F. Claerbout, 'Toward a unified theory of reflector mapping' *Geophysics*, 36, (1971), pp. 467-481.

Fundamentals of geophysical data processing. (McGraw Hill, 1976).

J.F. Claerbout and E.A. Robinson 'The error in least-squares inverse filtering' *Geophysics*, 29, (1964), pp. 118-120.

J.F. Claerbout and A.G. Johnson, 'Extrapolation of time dependent waveforms along their path of propagation' *Geophys J. of the Roy. Astr. Soc.*, 26, (1971), No. 1-4, pp. 285-294.

G.K.C. Clarke, 'Time-varying deconvolution filters' *Geophysics*, 33, (1968), pp. 936-44.

D.A. Disher and P.J. Naquin, 'Statistical automatic statics analysis' *Geophysics*, 33, (1970), pp. 574-85.

J.W. Dunkin and F.K. Levin, 'Effect of normal moveout on a seismic pulse' *Geophysics*, 38, (1973), pp. 635-42.

I. Finetti, R. Nicolich and S. Sancin, 'Review on the basic theoretical assumptions in seismic digital filtering' *Geophys. Prosp.* 19, (1971), pp. 292-320.

E.A. Flinn, (editor), E.A. Robinson and S. Treitel, 'Special issue on the MIT Geophysical Analysis Group reports' *Geophysics*, 32, (1967), pp.411-525.

P. Hubral, 'Time migration — some ray theoretical aspects. *Geophys. Prosp'*. 25, (1977), pp. 738-45.

G. Kunetz and J.M. Fourmann 'Efficient deconvolution of marine seismic records' *Geophysics*, 33, (1968), pp. 330-38.

Loewenthal, Lu, Robertson and Sherwood. 'The wave equation applied to migration and water bottom multiples' (S.E.G., Dallas, Nov. 1974).

H.J. Meyerhoff 'Horizontal stacking and multichannel filtering applied to common depth point seismic data' *Geophys Prosp.*, 14, (1966), pp. 441-54.

H.J. Neidell and M.T. Taner, 'Semblance and other coherency measures for multichannel data' *Geophysics*, 34, (1971), pp. 482-97.

K.L. Peacock and S. Treitel, 'Predictive deconvolution: theory and practice' *Geophysics*, 34, (1969), pp. 155-69.

R.A. Peterson and M.B. Dobrin, *A pictorial digital atlas*. United Geophysical, Pasadena, 1966).

E.A. Robinson, 'Multichannel Z-transforms and minimum-delay' *Geophysics*, 31, (1966), pp. 482-500.

Multichannel time series analysis with digital computer programs. (Holden-Day, San Francisco, 1967).

E.A. Robinson and S. Treitel, 'Principles of digital filtering' *Geophysics*, 29, (1964), pp. 395-404.

'Principles of digital Wiener filtering' *Geophys. Prosp.*, 15, (1967), pp. 311-33.

The Robinson-Treitel reader. Seismograph Service, Tulsa, 1973).

J.C. Robinson, 'Computer-designed Wiener filters for seismic data' *Geophysics*, 37, (1972), pp. 235-59.

R.E. Sheriff, *Encyclopaedia dictionary of exploration geophysics*.(Soc. of Exploration Geophysicists, Tulsa, 1973).

J.W.C. Sherwood and P.H. Poe, 'Continuous velocity estimation and seismic wavelet processing' *Geophysics*, 37, (1972), pp. 769-87.

M.M. Slotnick, *Lessons in seismic computing*. (Soc. of Exploration Geophysicists, Tulsa, 1959).

W.M. Telford, L.P. Geldart, R.E. Sheriff and D.A. Keys, *Applied geophysics*. (Cambridge University Press, Cambridge, 1975).

S. Treitel, 'Principles of digital multichannel filtering' *Geophysics*, 35, (1970), pp. 785-811.

S. Treitel and E.A. Robinson, 'Seismic wave propagation in terms of communication theory' *Geophysics*, 31, (1966), pp. 17-32.

'The design of high-resolution digital filters' *IEE Trans. Geoscience Electronics*, GE-4 (1966), **pp.** 25-38.

G.P. Wadsworth, E.A. Robinson, J.G. Bryan and P.M. Hurley, 'Detection of reflections on seismic records by linear operations' *Geophysics*, 18, (1953), pp. 539-586.

R.J. Wang and S. Treitel, 'Adaptive signal processing through stochastic approximation' *Geophys. Prosp.*, 19, (1971), pp. 718-28.

R.A. Wiggins and E.A. Robinson, 'Recursive solution to the multichannel filtering problem' *J. Geophys. Res*, 70, (1965), pp. 1885-91.

4. WELL DATA

When the interpreter comes to establish a tie between his seismic sections and a borehole section he faces the problem of making a direct correlation between patterns of reflectors which are scaled vertically in terms of two-way reflection time and the realities of sub-surface geology as determined by lithological logging of rock chippings and cores obtained from a borehole. The geologist's lithological log is of prime importance in that it provides the basis for identification of reflectors in terms of boundaries between rocks of different type. Other geological work on the cores and chippings aims to establish the age and stratigraphy of the geological section and the presented results of exploration drilling normally include a litho-stratigraphic log (rocks described in terms of lithology) as well as a chrono-stratigraphic scale (the rock units subdivided according to age).

It is standard industry practice that at various stages during the drilling of a well and upon reaching total depth (referred to on logs as TD) geophysical logging tests are made with a variety of instruments. These are lowered to the bottom of the well, as drilled at the time of logging, on a wire line which is usually a multi-core electrical cable on which the logging tools can be suspended. The logging tools are then drawn upwards through the borehole, measurements of various parameters being made either continuously or by tests at selected horizons. The processed results of these geophysical tests provide data which allow identification of the inter-relation between the seismic section time scale and the borehole section depth scale and thereby direct correlation between reflector pattern and stratigraphy. These measurements also provide data on the physical properties of the rocks penetrated by the borehole and such data are important to a geological understanding of the variation in reflector pattern which can be seen in seismic sections throughout an exploration province. As far as the seismic interpreter is concerned, the geophysical logging methods of most value to him are gamma-ray logging, compensated sonic logging and well velocity surveys. The results of these are most usefully combined to provide a synthetic seismogram which is a process which aims to produce from the borehole physical data a computed seismic section display which should compare well with an actual seismic section surveyed through the well site.

4.1 Logging tools

Wherever possible during logging, a number of individual tools are assembled together so that each logging run will allow concurrent acquisition of data for a group of parameters. Thus the results of most logging runs are displayed as a multi-trace vertical profile showing the measurements pertaining to the tool assemblage used (see for example, figure 4/2 which shows a display of borehole diameter as

measured by a caliper log, a gamma-ray log and a bulk density log).

4.1.1 Caliper log

Because the variations in borehole diameter must be accounted for in data processing of the results of most geophysical tests in open boreholes, a device is attached to the logging tool assemblage which measures variation in the width of the hole throughout the interval logged. Results of a caliper logging run can often provide a useful indication of lithological variation and these are used in the geological interpretation of any suite of logging tests.

4.1.2 Gamma-ray log

The gamma-ray logging tool measures the natural radioactivity of the formations in the borehole and it is used as a means of identification of shales, some evaporites and of igneous rocks and igneously derived sediments. An important feature of the log is that it can be used in a cased borehole and can substitute for electrical logs as a means of correlation between boreholes. Separate runs in the same borehole can be seen in figures 4/2 and 4/5. This log is usually only used as an aid to seismic correlation if sonic and density logging data are not available.

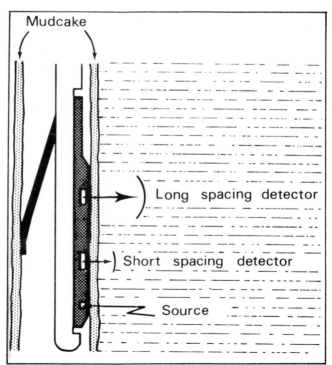

Figure 4/1: Formation density logging tool. (*Courtesy: Schlumberger*).

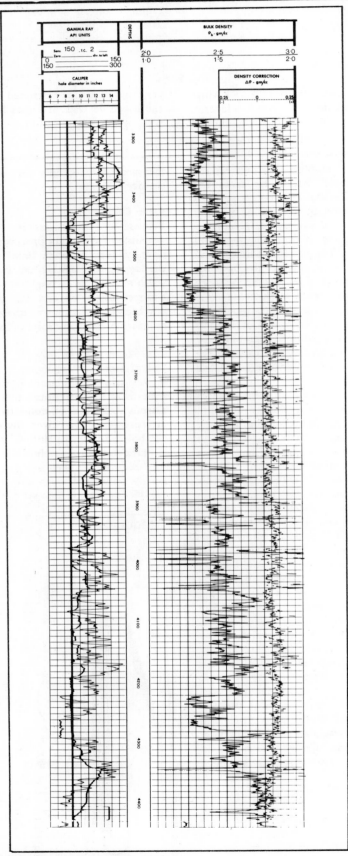

Figure 4/2: Formation density log, Western Canada. Bulk density and density correction are displayed to the right of the depth scale; gamma and caliper logs to the left. Note the extensive caving as shown on the caliper log. In acoustic studies, corrections for caving would be necessary.

4.1.3 Compensated formation density log

The formation density logging tool is constructed as shown in figure 4/1. The instrument is an arrangement of a gamma-ray source and two shielded detectors which measure the intensity of gamma-ray back scatter from the rock formation. The amount of back scatter is approximately proportional to the electron density in the rock which is proportional to the rocks bulk density. Because a layer of mud-cake commonly forms on the wall of the borehole it is necessary to compensate for its effect, hence the use of two detectors as shown in figure 4/1. Natural radioactivity of rocks does not affect the results obtained, but, unlike the gamma logging tool, this log can only be successfully used in uncased holes.

In figure 4/2 an example is shown of a log from western Canada with a range of rock types from Lower Cretaceous coals, density 1.35g/cc, to carbonates with a density of 2.8g/cc. Like the sonic logs (see below) the density log is invaluable to the reservoir geologist in determining porosity variations. This is due to the fact that there is a large density contrast between rock matrix material and pore-space fluids. When combined with a neutron density log (not described here) it provides a means of detecting gas accumulations.

A recently developed alternative method of measuring density variations in rocks by use of a logging tool is through use of a borehole gravity meter. Use of this tool requires the determination of gravity to a very high accuracy at selected locations in the hole and at present tests are usually undertaken as a means of studying prospective intervals rather than as a means of obtaining a continuous profile of density variation throughout a deep well. However, the results are reliable, accurate, and the zone tested is not restricted to the immediate vicinity of the borehole wall. It seems likely therefore that this tool will have wide application in the future.

4.1.4 Compensated sonic log

The sonic logging tool measures the reciprocal of velocity in rocks of the walls of a borehole. As shown in figure 4/3 it consists of a pair of units each with a sound generator and two receivers. Pulses of sound are refracted along the wall of the borehole and the time differences are measured between receipt of these pulses at each of the group of two receivers. The time differences are divided by the distance between the receivers to give a transit time in microseconds per foot. The two units pulse separately and the average of the values observed is used to compensate for errors caused by irregular borehole configuration or angular deviation of the instrument. The log is normally integrated to give total travel time on a scale of milliseconds and interval times may be obtained by counting scaled interval units (see figure 4/4). Interval velocities are obtained by determining the reciprocal of the transit time. It should be noted that the instrument occasionally skips a cycle due to non-arrival at a far receiver or a late arrival at a near receiver, the transit times being respectively too long or too short. This can be readily recognised by the sudden deflection on the curve and compensated for manually. This logging tool can only be run in an open, uncased hole.

An example from western Canada is shown in figure 4/5. Rock types range from coals (130 microsec/ft or 2.35km/s) through shales and sands to carbonates (50 microsec/ft or 6.1km/s). Like the density log this log can be used by the reservoir geologist as a porosity tool. This depends on the fact that there is a large velocity difference between rock matrix materials and pore fluids.

4.1.5 Well velocity surveys

A well velocity survey is the most direct method of identifying the relationship between sub-surface geology and seismic reflection data. The technique involves detecting sound from a near surface source with a pressure geophone (hydrophone) at selected levels within the fluid-filled borehole. These levels are usually chosen with reference to major changes in formations in the geological section. In offshore wells the source is usually an air gun whereas on land dynamite in shot holes may be used or an air gun submerged in a water-filled hole. A complete set of equipment is illustrated in figure 4/6.

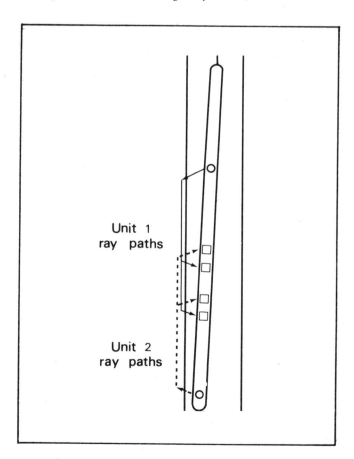

Unit 1
ray paths

Unit 2
ray paths

Figure 4/3: Sonic logging tool.
(*Courtesy: Schlumberger*).

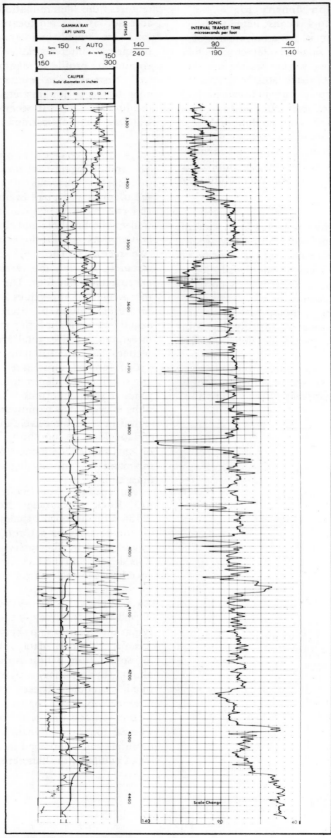

Figure 4/5: Borehole compensated sonic log, Western Canada. This log covers the same interval as in figure 4/2. Note the change in borehole diameter between logging runs as indicated by caliper log. CVL values range from 130 to 50 μsec/ft (2.35 to 6.1km/s) reflecting a range of rock types from coal, through shales and sandstones to carbonates.

km/s	2·18	3·39	7·26
ft/s	7,143	11,111	25,000
μsec/ft	140	90	40

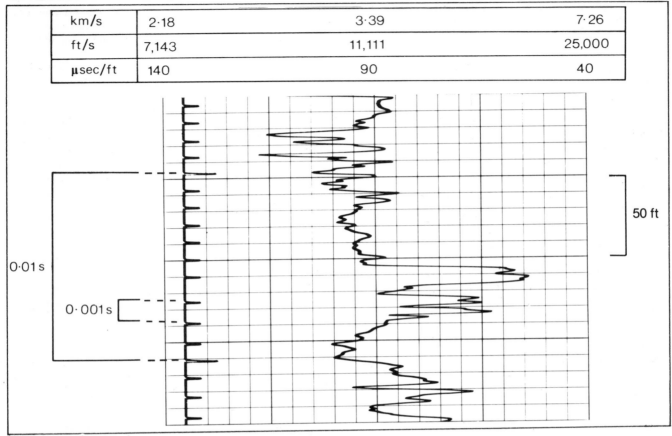

Figure 4/4: Derivation of integrated total travel time from continuous velocity log (CVL). On the right is shown part of a normal CVL log and on the left is displayed the integrated total travel time for this part of the section. By totalling from the surface downwards the two-way travel time to a marker horizon can be calculated. Interval velocities can also be derived.

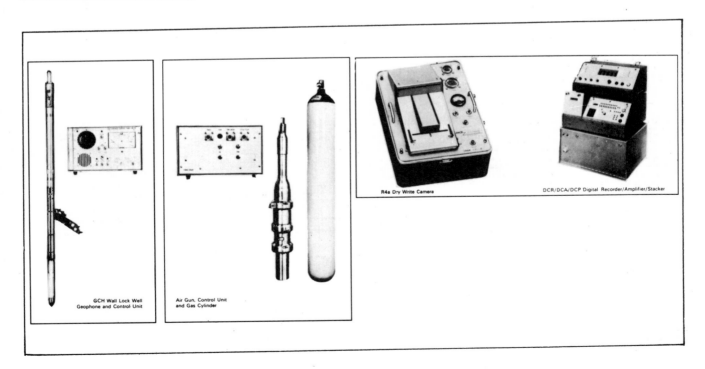

Figure 4/6: Well velocity survey equipment.
(*Courtesy: Seismograph Service Ltd (SSL)*).

When surveys are being carried out it is important that certain requirements are met of which the most important are as follows:

1. The source must be a sufficient distance away from the hole, proportional to the expected test level depths, to avoid refracted first arrivals via the casing.

2. If it is desired to make a test close to an important stratigraphic boundary, the hydrophone should be located at approximately 5 metres below that boundary to avoid adverse effects of poor hole conditions such as caving or loose fill in the vicinity of the test.

3. For each measurement the logging tool must be locked and the wire line slackened off to avoid cable noise.

An annotated monitor record is shown in figure 4/7.

Data reduction includes the correction of the recorded times to true vertical and these can then be used to calibrate a sonic log. Corrections are by linear or differential shift, whichever is most appropriate. The final presentation is as a calibrated acoustic log which can be directly overlain on a seismic section which intersects the borehole. An example of such a calibrated log is shown in figure 4/8.

Through use of digital acquisition equipment it is possible to derive additional data from well velocity surveys than that required to produce a calibrated sonic log. The data for each test level provide a record which is equivalent to a reflection seismic trace with a deeply buried detector and events other than first arrivals can be processed using methods similar to those used in processing conventional reflection data. Because the detector is buried, both upward and downward travelling waveforms will be recorded from reflecting horizons above and below the hydrophone location as well as multiples generated in the succession. Techniques are available which can remove the unwanted reverberant events and the product can be displayed in a form similar to that of a variable-area seismic section as a vertical seismic profile (VSP). In figure 4/9 is shown a monitor record on which secondary events can be clearly recognised. This figure shows also how such records can be processed and stacked to give a section illustrating the major reflectors even at depths below the TD level of the borehole.

In most circumstances the well velocity survey is the last logging run made before production casing is fitted or before the well is plugged and abandoned. In such circumstances there will always be pressure to complete the work as quickly as possible. Nevertheless, the value of a good well survey to future interpretation work in a province cannot be over-emphasised and every precaution should be taken to ensure that an adequate number of levels have been tested and that where necessary several shots have been recorded at each level depending on data quality.

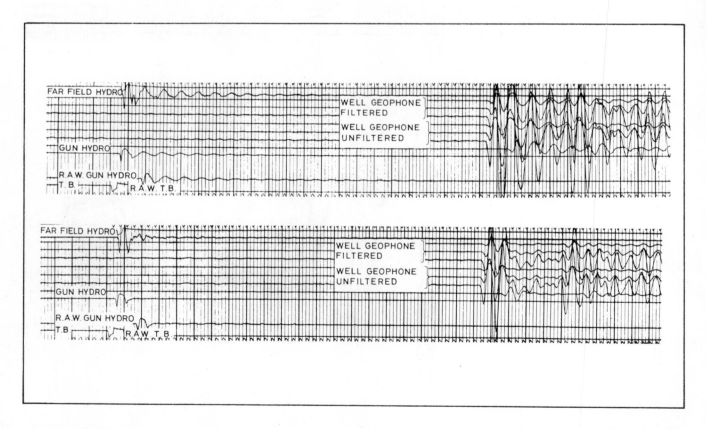

Figure 4/7: Velocity survey field monitor record. These records display the electronic time break, the signal from a gun hydrophone which monitors both the time and waveform of the emission from the gun as well as the downhole detector signal displayed on six traces on which the gain and filtering characteristics can be varied. (*Courtesy: SSL*).

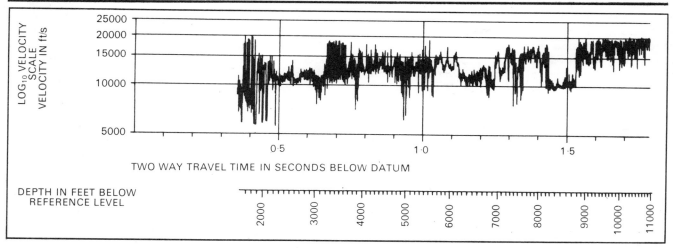

LOG₁₀ VELOCITY SCALE VELOCITY IN ft/s

LOG_{10} VELOCITY SCALE VELOCITY IN ft/s

TWO WAY TRAVEL TIME IN SECONDS BELOW DATUM

DEPTH IN FEET BELOW REFERENCE LEVEL

Figure 4/8: A calibrated sonic log. Measured time intervals from the velocity survey are used to correct and calibrate integrated travel times from the sonic log. The sonic log is then redisplayed against a vertical linear two-way travel time scale to match seismic sections through or adjacent to the well. A non-linear depth scale is also displayed.
(*Courtesy: SSL*).

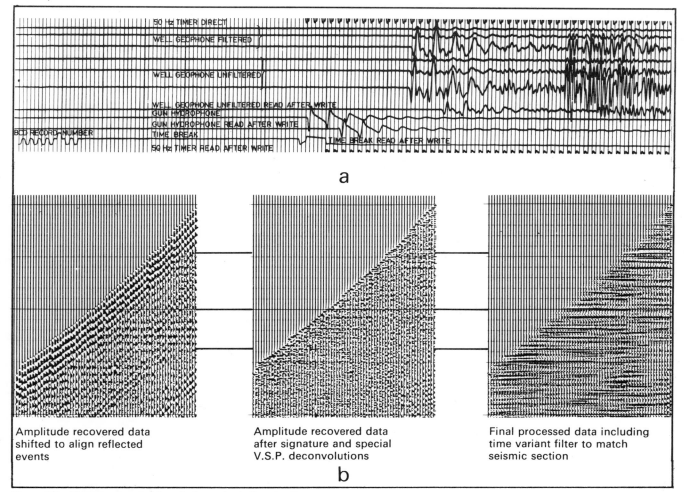

Amplitude recovered data shifted to align reflected events

Amplitude recovered data after signature and special V.S.P. deconvolutions

Final processed data including time variant filter to match seismic section

Figure 4/9: Monitor record (above) shows both first break event and secondary events. Secondary events can represent both upward and downward travelling wave trains and can originate from reflectors deeper than the borehole hydrophone. Digitally processed records can be stacked and displayed to match adjacent seismic sections. (*Courtesy: SSL*).

4.2 Synthetic seismograms

Well velocity survey results are usually the last item of borehole information to be delivered, often a considerable time after abandonment of drilling. In view of the ever-changing priorities of an exploration department, a sufficient amount of tying-in will have been performed by then. Nevertheless, the synthetic seismogram can be of great value to the interpreter and it is best presented by splicing it to an interpreted seismic section through the well location.

The process of production of a synthetic seismogram is illustrated in figure 4/10. It is prepared by displaying the convolution product of the reflectivity coefficient (as discussed in chapter 1) with a suitable zero or minimum phase wavelet. Care should be taken that the frequency response should be the same as that of records in the vicinity. Zero minimum phase or Ricker wavelets are currently in fashion. The velocities for the reflection coefficient are obtained from the calibrated (if well velocity survey is available) integrated sonic log and densities from the formation density log. Use of densities in deriving the reflection coefficients is optional and sometimes omitted to save processing time and costs. Often the difference is minimal but for high resolution work use of both variables is recommended. Many land wells are completed without obtaining a well velocity survey. As a consequence, most land synthetic seismograms mis-tie to some degree due to the absence of calibrating check shooting. It should be noted that the synthetic seismogram can be constructed manually with a little time and care using only graph paper and a pocket calculator and an interpreter may find this an ideal way of checking thin-bed tuning and possible

hydrocarbon indicators. The synthetics shown in figure 4/10 illustrate the range of types and displays available. Standard displays include: primaries only, primaries without transmission loss, all order multiples, primaries plus all order multiples. Often it is necessary as a means of obtaining a better match to nearby reflection seismic data to obtain extra optional displays which may include: primaries plus short-term multiples for any given time zone width, primaries plus ghosts, primaries plus all order multiples plus ghosts.

4.3 The composite log

The interpreter who utilises the results of well logs and surveys is heavily dependent on geological identification as well as age dating. Before a final composite log is produced, geological studies are necessary and in particular by way of undertaking lithological and stratigraphic correlations from drill cuttings and cores by comparison with other wells. Also, proper geological identification will require a range of palaeontological, micropalaeontological and/or palynological studies of the core and chipping samples. Geophysical logs will be used to estimate formation correlation between wells by comparison of, for example, sonic and gamma-ray logs. The final geological analysis of a borehole is detailed in a composite log which usually consists of:

1. A lithological description including palaeontological information.
2. A borehole compensated sonic log.
3. A gamma-ray log.
4. A caliper log.
5. Formation identification.

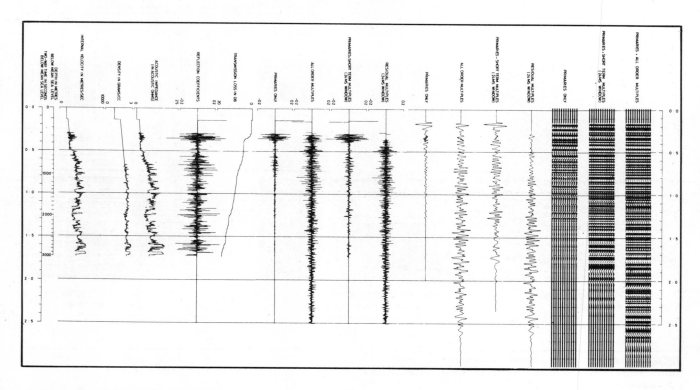

Figure 4/10: Range of synthetic seismogram displays derived from density and interval velocity logs.
(*Courtesy:* SSL).

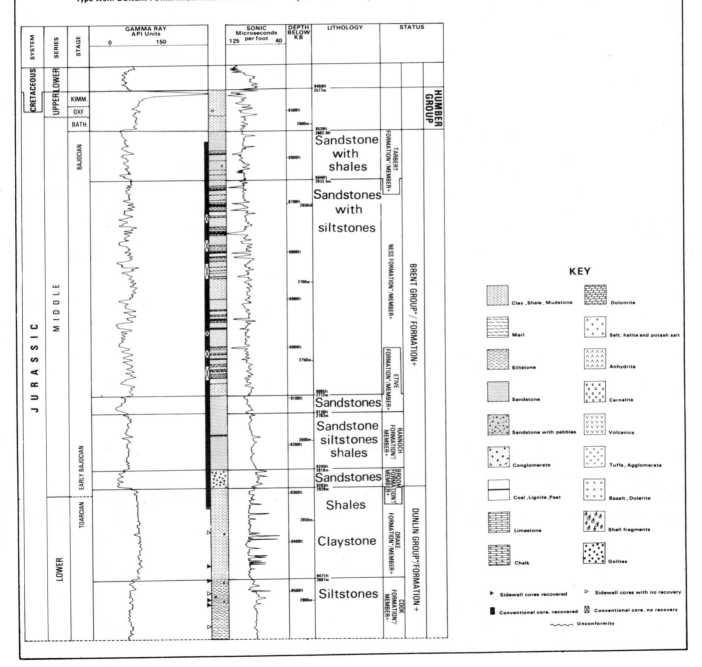

JURASSIC
UK WELL 211/29-3 (BRENT FIELD)

OPERATOR: SHELL
CO-ORDINATES: 61°08′06″N 01°43′36.5″E
SPUDDED: 14/7/73   DRILLING COMPLETED: 14/10/73
KBE: 24m (78ft) WATER DEPTH: 177m (580ft)

Type Well: AMUNDSEN, BURTON, COOK AND DRAKE FORMATIONS (DUNLIN GROUP),
AND BROOM, RANNOCH, ETIVE, NESS AND TARBERT FORMATIONS (BRENT GROUP) *[U.K. USAGE]*

Type Well: DUNLIN FORMATION AND BRENT FORMATION *[NORWEGIAN USAGE]*

Figure 4/11: North Sea composite log (after C.E. Deegan and B.J. Scull, (compilers). 'A proposed standard lithostratigraphic nomenclature for the Central and Northern North Sea'. *Rep. Inst. Geol. Sci.*, No.77/25; *Bull. Norw. Petrol. Direct.*, No.1. (1977).

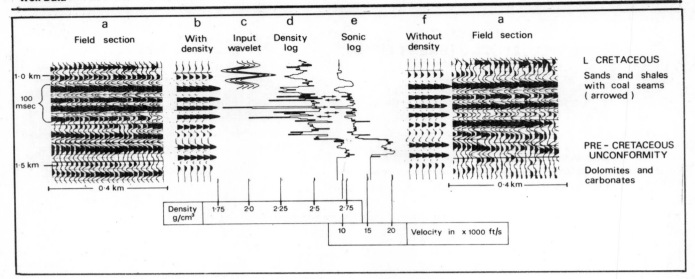

Figure 4/12: Example of borehole tie from Western Canada. The density and sonic logs (d) and (e) are rescaled versions of the logs shown in figures 4/2 and 4/5. Synthetic seismograms (b) and (f) are prepared by convolving the input wavelet (c) with reflections derived from the geophysical logs. Displays (b) to (f) are spliced into the field section (a). The correspondence between field sections and synthetic sections is good. A group of reflections can be seen to be associated with the coal seams and it should be noted that although in some cases reflectivities are doubled when calculated using density information (coal has a very low comparative density), the synthetics show only minor amplitude variation because of the thinness (5—17ft) of the seams.
(*Courtesy: Digitech Ltd*).

6. Information on drilling rate (optional).
7. Information on drilling mud weight (optional).

Where the well has penetrated and/or detected hydrocarbons, pertinent data will be listed which may be utilised by the geophysicist in seismic hydrocarbon indicator studies. A typical composite log from the North Sea is shown in figure 4/11.

4.4 Seismic ties to borehole data

If a seismic section can be tied directly to well information, the synthetic seismogram (with the correct polarity and frequency band-width) should be overlain or spliced in at the appropriate location. As discussed earlier in the section on seismic pulses and the earth as a filter, it is unreasonable to expect a perfect match in amplitude, frequency and phase. Where there is a good fit, often a time mis-tie will be found due to incorrect NMO and/or static correction, incorrect corrections for weathering or other low velocity datum corrections on land, and at sea, the possible use of incorrect datum corrections for gun and cable depths. Other mis-tie effects may be induced by phase distortion in the recording or playback instrumentation. Correlation should therefore be made on an interval best-fit basis as has to be done with the synthetic seismogram which has been produced from an uncalibrated sonic log. Static errors should always be investigated but may be unresolved.

In the absence of, or in conjunction with a synthetic seismogram, the log velocity scale from a well velocity.

survey or sonic can be overlain or spliced into a seismic section and velocity contrasts aligned with appropriate peaks . and troughs (see figure 4/12). Where the above are not available, interval velocities can be plotted on a suitable time scale from a continuous velocity log and compared as above. If neither synthetics nor well velocity surveys are available, the integrations on a sonic log can be plotted manually as an interval velocity curve and used similarly. If no sonic logs are available, a formation density log can be used to give a very qualitative indication of the relationship between the geological section penetrated by a well and the equivalent seismic section. If there is some confidence in the velocities derived regionally from seismic data, the gamma-ray or formation density depth scale can be converted to time and empirical correlations made.

When seismic lines do not tie directly to boreholes, various approaches can be adopted for reflection identification but mainly in this situation there is a considerable reliance on intuition and there can be no substitute for a good borehole tie.

References and suggested reading

V. Baranov and G. Kunetz, 'Film synthetique avec reflexions multiples theorie et calcul pratique' *Geophys. Prospect.* 8, (1960) pp. 315-25.

H. Guyod and L. E. Shane, *Geophysical well logging,* Vol.1, *Introduction to acoustical logging.* (H. Guyod, Houston, 1969).

L.G Howell, K.O. Heintz and A. Barry 'The development and use of a high precision downhole gravity meter' *Geophysics* 31, (1966) pp. 764-72.

P. Kennett and R.L. Ireson, 'Recent development in well velocity surveys and the use of calibrated acoustic logs' *Geophys. Prospect.,* 19, (1971) pp. 395-411.

F.P. Kokesh and R.B. Blizard, 'Geometric factors in sonic logging' *Geophysics,* 24, (1959) No. 1 (February).

F.P. Kokesh, R.J. Schwartz, W.B. Wall and R.L. Morris, 'A new approach to sonic logging and other acoustic measurements' *Journ. Pet. Tech.,* 17, (1965) No. 3 (March).

R.M. Pegrum, G. Rees and D. Naylor *Geology of the North West European Continental Shelf* Vol. 2. (Graham and Trotman Ltd., London, 1975).
This book is mainly concerned with the hydrocarbon geology of the North Sea, but exploration techniques, including well logging are discussed.

R.A. Peterson, 'Synthesis of seismograms from well log data' *Geophysics,* 20 (1955) pp. 516-38.

Schlumberger Ltd. *Log interpretation principles* (Schlumberger, New York, 1972).

Seismograph Service Ltd. *Air-gun land and marine well velocity surveys.* (SSL, Holwood, U.K. 1976).

J. Tittman and J.S. Wahl, 'The physical foundations of formation density logging (Gamma-gamma).' *Geophysics,* 30 (1965) April.

M.P. Tixier, R.P. Alger and D.R. Tanguy, 'New developments in induction and sonic logging' *Journ. Pet. Tech.,* 12, (1960), No. 5 (May).

J.S. Wahl, J. Tittman and C.W. Johnstone, 'The dual-spacing formation density log' *Journ. Pet. Tech.,* 16 (1964) December.

M.R.J. Wyllie, A.R. Gregory and G.H.F. Gardner,'An experimental investigation of factors affecting elastic wave velocities in porous media' *Geophysics,* 23 (1958), No. 3 (July).

5. GEOPHYSICAL INTERPRETATION

The object of geophysical interpretation is usually to prepare contour maps showing the depth to a series of reflectors which have been picked on the seismic sections. The work falls into several parts which are described below.

5.1 Quality control of survey and processing

The geophysicist will often be involved in a seismic survey from its planning to its interpretation. It is therefore appropriate to consider here the input that the interpreter can make before the seismic sections arrive on his desk.

In planning a survey grid, it is important to think carefully about the purpose for which the data is required. In the case of a regional survey, a rectangular grid is often best, preferably orientated parallel to regional dip and strike, with dip lines being the more closely spaced. Line spacing may be dictated by constraints on overall cost, but should be at least as close as the distance between major features of the area (for example, major faults or salt structures). Lines through boreholes are especially desirable, since they will often give a firm indication of reflector identity. Surveys intended to solve some more specific geological problems are generally made at a stage when a reasonable amount of data is already available, so that selection of a survey pattern will depend on the problem and the pre-existing data. Other parameters to be specified include the total record length, which depends obviously on the time to the deepest expected event, and the sampling interval, generally taken as 4msec but occasionally as small as 1msec if a particular geological situation requires high resolution, for example, for reservoir delineation. Specifications for maximum permissible ambient noise, source misfire rate, number of geophones malfunctioning, feathering angle of cable (at sea), and deviation from intended survey line are fairly standardised, but it is worth remembering that tighter limits may have to be applied if true-amplitude processing is contemplated. During the survey, the geophysicist has a role as client representative either on board the seismic ship, or with the land crew, monitoring progress and ensuring that specifications are met.

Liaison between the interpreter and the processor of seismic data is of great value. Much of the interpretation of the data is implicitly carried out at the processing stage. This is especially true of the velocity analysis stage where events can be enhanced or made to disappear completely in the subsequent stack. Geological knowledge of the area is therefore of great value, both in checking suggested velocity distributions for plausibility (especially in the presence of anomalies such as velocity inversions) and in indicating which events are of most importance to the understanding of the particular area, for it is to these that particular attention should be given. Comments on filter settings,

deconvolution tests, and final display mode can all be very valuable. Specification of special processing required, such as migration, can also often be made at this stage. Quality control consists not only of checking for consistency of reflectors at line intersections, and of velocities along lines, but generally making sure that any anomalous features of the sections are not artefacts of processing, such as, for example, a sudden change of processing parameters without geological justification.

5.2 Picking a survey: reflection identification

It is usually best to start picking a survey by inspecting lines through boreholes. Not only do the well logs give a useful geological picture, but they also show where strong reflections might be expected. It is sufficient for a first look to use the sonic log anticipating an event at each major seismic velocity change, provided that the bed giving rise to it is at lease one wavelength (say 100m) thick. It should however be noted that it is possible for in some circumstances quite thin beds to produce strong reflections, down to one quarter wavelength thickness or less. This can be due, at a particular frequency, to constructive interference between reflections from the top and base of the bed. Such a reflector may be locally of great importance, because it may make possible detailed study of a hydrocarbon zone. On a broad scale, such reflectors are apt to have a very variable character and can disappear as bed thickness changes laterally. Sometimes, however, density varies in the opposite sense to velocity and changes in acoustic impedance, and therefore reflection strengths are low despite large velocity changes; this is often the case with salt layers. If there is any doubt at all over reflection identity, it is necessary to use a time-depth log incorporating the results of well check shots, if one exists, to relate the seismic events to the geological horizons. If there is no such log various approximate methods can be used (see p.67). For an example of such a borehole tie, see the Moray Firth Case Study (p.125).

The next stage is to examine briefly the entire survey. Often it is best to start with the dip lines which are usually easier to interpret, and lay them out in sequence. In this way it is possible to follow the major structures across the area, and at this stage it is worth marking the main faults, treating apparently major structures visible on only one line with some suspicion. Although reflector discontinuity indicates the presence of a fault, diffraction patterns are often useful in locating the fault plane precisely; the plane should pass through the apex of the hyperbola. There is often a tendency for reflectors to continue for a short distance across a fault plane owing to the effects of diffraction described in chapter 1; the resulting appearance resembles a thrust fault (figure 5/1). Caution should therefore be exercised in marking reverse-throw faults on seismic sections unless independent evidence for their existence is known.

It is now possible to mark on the section those reflectors whose identity is known from boreholes. Fine-pointed coloured pencils are a convenient method, since a uniform colour code can be established throughout an area. For each reflector, the first peak should be marked. Usually, faulting or other structure makes the following of reflectors difficult within a short range of a borehole, since it will commonly have been drilled on a structural high within a

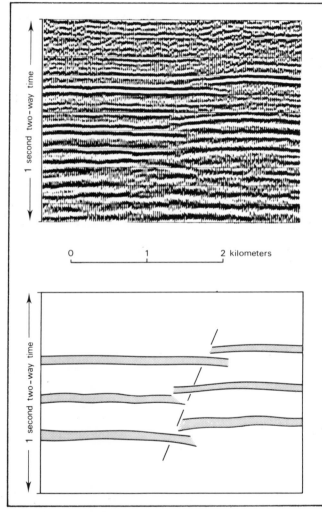

Figure 5/1: Interpreted seismic section showing typical normal faulting.
(*Courtesy: IGS; Seiscom survey*).

basin, and therefore in an area of tectonic disturbance. The problem of correlation across such structures thus arises. It is often possible to correlate across small faults by using strike lines to carry the interpretation around the ends of them; in the case of major structures it is usual, in the absence of any other information, to use the reflection character as a guide. It can be helpful to fold a section so as to bring into juxtaposition the undisturbed area on the two sides of the fault, when the correct correlation is sometimes obvious; doubtful cases should however be treated with suspicion since the character of a reflector is highly dependent on processing parameters which may change

greatly over a major fault. Geological plausibility is also a useful guide; faults whose throws diminish markedly downwards should cause reconsideration, although such effects are geologically possible and are not uncommon on seismic sections where velocities generally increase with depth.

Finally, the picking of the entire survey should be tied together, making sure that all line intersections are consistent. This is a very powerful check on the correctness of the interpretation, but some problems often occur. It is often useful to make a rough map of the main structural features at this stage. Figure 5/2 shows a fictitious example of such a map. In figure 5/3 the intersection ties from a closed loop are illustrated.

5.3 Closing loops: mis-ties and their causes

It is sometimes necessary to introduce faults unsupported by other evidence in order to make sure that a picked reflector ties round a loop of survey lines. Before such drastic action, however, other causes of mis-tie should be considered. Assuming that the picking has been checked around the loop and no plausible adjustment to secure a tie can be made, we must consider the possible causes in the survey and its processing for such a difficulty.

Especially with older marine data, shot before widespread use of satellite navigation systems, poor navigation can be a cause of gross mis-ties. Water depth, if written on the section as it usually is, can indicate the presence of a navigational problem and suggest how the positions of the lines should be adjusted to achieve a tie. If a line crossing a grid survey requires a consistent position adjustment in a particular direction, the interpreter can have considerable confidence in laying the blame for mis-ties on faulty navigation; if the adjustments required are random in direction and magnitude, other possible causes must be examined.

A common cause of difficulty is a reversal of polarity between sections. Again, particularly with older data, the polarity may be unknown or even incorrectly stated. It is often difficult to know whether a polarity reversal is present or not if it is combined with shifts due to other causes. A reflector of characteristic appearance will provide a test, but one must remember that the appearance of an event can be drastically changed by differences in processing parameters, especially in stacking velocities, deconvolution parameters, and filter settings; this change in appearance is itself a frequent cause of mis-ties. There is usually a consistent mis-tie of about 40msec between 'Maxipulse' data and data from other sources; generally the 'Maxipulse' data ties into wells more accurately. Other causes of mis-ties are differences in stacking velocities (due not only to mispicking of velocity displays but also to interpolation between velocity analyses causing a mis-tie of stacking velocities at an intersection), and the effect of dip which may cause the actual reflection points at the intersection of two unmigrated sections to be quite different on the two lines.

In practice if there is an appreciable non-navigational mis-tie between surveys it may be necessary to examine all the intersections to establish an average shift to be applied to all the sections of one survey. If the shift required is small it may often be neglected in regional mapping.

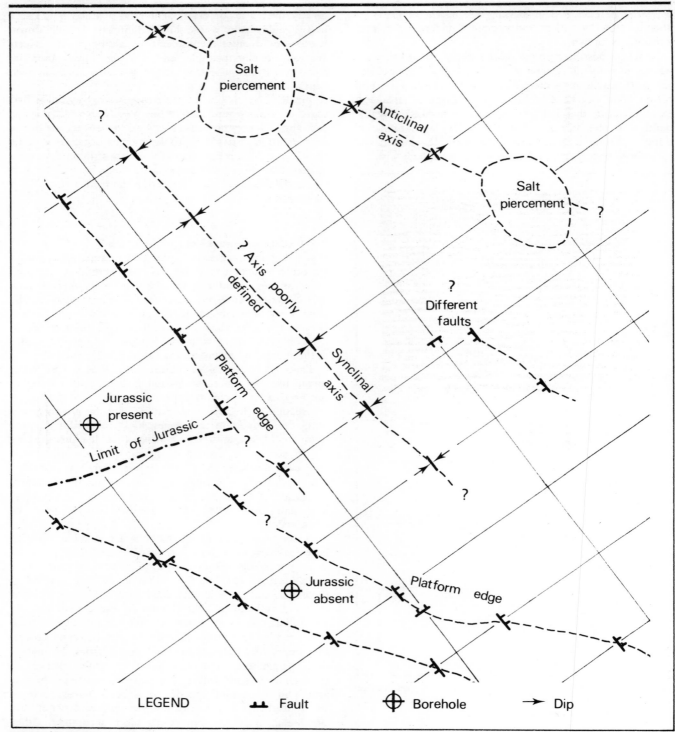

Figure 5/2: A schematic structural map of a seismic interpretation project area.

5.4 Digitisation

The first step in making a map from the picked seismic sections is the measurement of two-way time to the picked events along each section. These data will then be transferred to the base-map in order to produce a contour map of structure in two-way time.

The simplest method of reading the times off the sections is to perform the operation directly by hand, using a scale appropriate to the section. The horizontal intervals between readings depends on the complexity of structure

visible on the section; ideally, the frequency of reading should be such that interpolation between adjacent points by straight lines is adequately precise. Much depends on the scale of the final map; for example, in regional mapping at a scale of 1:100 000, the sections might be measured every 1km except in areas of greater complexity. If, however, the data will be mapped at a larger scale at a later date, it is probably worth taking this future requirement into account from the beginning. Positions of faults should be recorded

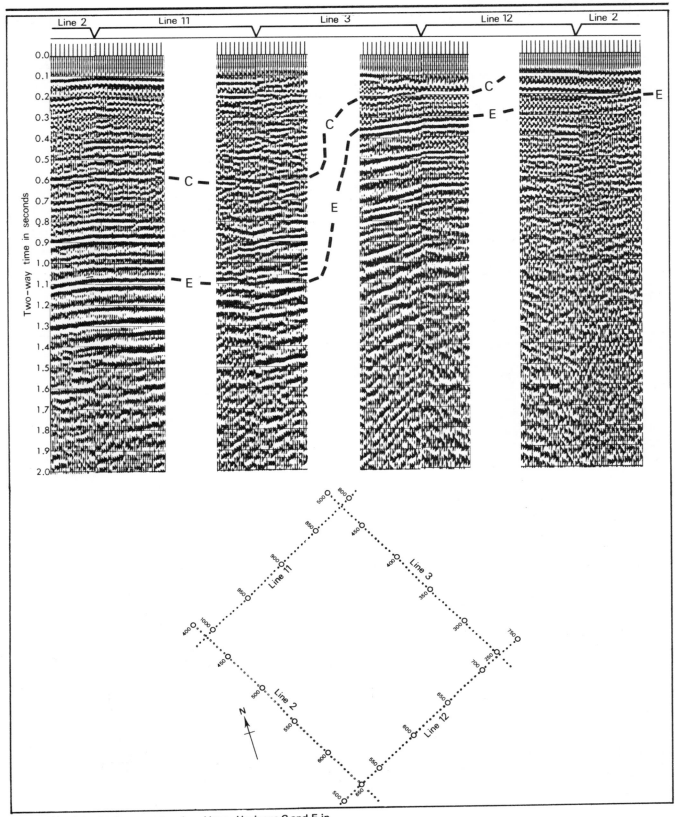

Figure 5/3: Seismic ties around a closed loop. Horizons C and E in the Moray Firth (*see chapter 9*) are shown at the intercepts of lines 2 and 11, 11 and 3, 3 and 12, 12 and 2 as shown on the shot-point map. A complete interpretation of lines 3 and 11 is shown in chapter 9. Horizon E can be traced continuously around the loop whereas horizon C is absent at the south end of line 2 and the west end of line 12. Note the change of reflector quality at the various intercept locations.

(*Courtesy: IGS data; Seiscom survey*).

and horizons timed on both the up and down thrown sides.

Reduction of data by hand in this way can become tedious if large amounts of data are involved. Systems exist for the direct digitisation of seismic records into computer-readable form, so that maps can be automatically prepared by machine. Typical digitiser hardware is illustrated in figure 5/4. A seismic section is fixed to the table and header information giving the line name, names of horizons to be digitised, etc, is typed in from a keyboard. A cursor is then moved along each horizon in turn and the (x,y) co-ordinates of the corners of the section are recorded so that the software may later calculate the scale factors in x and y directions and apply a correction for the skewness of the seismic record with respect to the (x, y) axis of the table. Output from the machine can be magnetic tape, paper tape, or cards; a preferable system is to connect the digitiser on-line to a computer, in which case data reduction can proceed simultaneously with digitisation, and errors can be seen and corrected before any incorrect information has been filed.

Various software packages exist for reduction of the digitised data to the equivalent of the manually-read values. Typically, data is scaled and converted to shot point versus two-way-time values, which are then interpolated to a uniform specified shot point increment. Static corrections, if specified in the header information, may be made automatically.

In general, data reduction by machine does not offer large savings in time unless a number of horizons are to be digitised on each section. This is because of the care required to enter correct header information, without which the subsequent reduction will be erroneous. Checking and correcting of the automatically prepared data files can also be time-consuming. The main advantages of having the raw data in machine-readable form are that updating of maps consequent on the receipt of new data is much simpler, and that depth conversion (see p.78) is facilitated. A flow chart for a typical system is shown in figure 5/5.

Figure 5/4: A digitising table with operator digitising a seismic section.
(IGS photo).

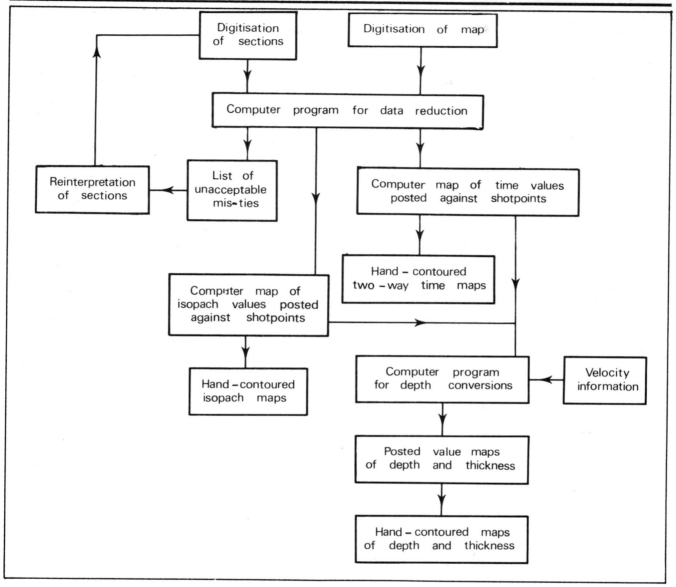

Figure 5/5: Flowchart of a scheme for digitising and depth converting the results of a seismic interpretation project.

5.5 Map construction

The first object is to produce a contour map of two-way-time to each horizon. Even where depth conversion is intended, this stage should not be omitted as the two-way-time map is directly related to the seismic sections and will need much less radical revision as new data is acquired than will the depth map, construction of which involves the use of (often very uncertain) velocity functions. If the two-way-times have been read off the section by hand, it is necessary next to post these on to a shot-point map. Before contouring the values it is desirable to mark all the faults on the map and decide how to join them together making sure that faults on maps of different horizons are coincident. Often it is necessary to refer back to the original sections to determine which faults are of similar appearance. Knowledge of dominant geological trends is helpful. Contouring is then carried out between the faults. Contour interval depends on the depth resolution required of the map, but a useful rule of thumb is that contours should

have an average spacing of about 1cm if the resulting map is to be both legible and in reasonable detail.

If the data have been digitised by machine, the next step is to digitise the shot-point map so that a map of two-way-times posted in their correct positions can be produced automatically. It is possible to proceed from here by computerised contouring, but few software packages available at present perform well in the presence of faults, and hand contouring is preferable.

It is difficult to give general advice on hand contouring, as most problems can be solved by common sense. Many people find it useful to begin by contouring the data roughly so as to identify the main trends, followed by a detailed revision; it can be useful to draw synclinal and anticlinal axes on the map so as to ensure that all contours intersecting them turn along the same axis. Minor mis-ties become apparent at this stage and can be resolved by reference to the original sections. Particularly in regional-

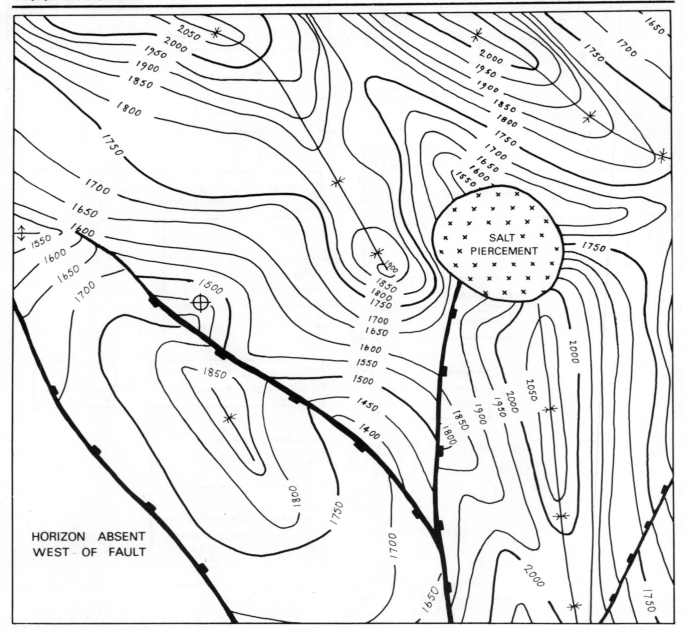

Figure 5/6: A typical two-way time map.

scale work, it is desirable to ignore an isolated value where it disturbs the trend.

When contouring has been completed, it should be checked against the original sections, especially in regions of complex structure (for example, in areas much disturbed by faulting) and in the vicinity of closed highs if they may be of hydrocarbon significance. The final map is then prepared; usually it shows faults of different throws by different widths of line (increasing in width as the throw becomes greater), and usually also some contours (say every fifth) are bolder than the rest to facilitate a quick appreciation of structure. A typical two-way time map is shown in figure 5/6. In figure 5/7 a typical mapping scheme is illustrated, this defines the symbols to be used, contour and fault symbols, line widths and style of presentation.

5.6 Velocity maps

In order to convert the two-way time map to a depth map, we need to know the velocity distribution over the area. The sources of information on velocities are well sonic logs on the one hand and seismic stacking velocities on the other. We shall consider each in turn.

In an area where drilling has been extensive but seismic data are sparse or of poor quality, velocities derived from seismics will be unreliable and it may be necessary to rely primarily on well velocity data. From the sonic log, the average velocity in each formation can be deduced. At first sight it seems that all that is required is to contour this information for each formation of interest. Unfortunately, however, wells are not normally scattered uniformly over an area; they are concentrated on the high areas within a

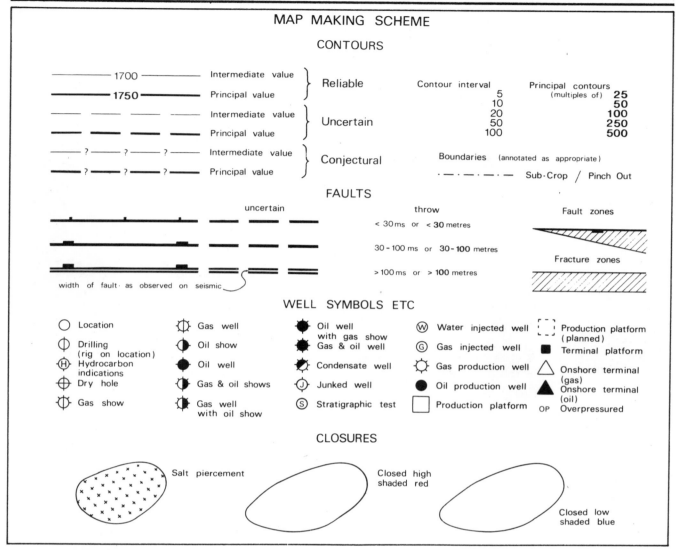

Figure 5/7: A mapping scheme with standard symbols used in oil industry mapping.

basin. The problem then arises as to how to extrapolate the well velocities into the deeper parts of the basin. One method is to measure the variation of velocity with depth of burial; in some cases a good correlation is obtained, but if an uplift subsequent to burial has occurred at some time the degree of recovery of compaction on release of overburden pressure will depend on the lithology. Therefore, a substantial scatter on a velocity-depth graph is common. If no other information is available, however, the best depth gradient of velocity may have to be used to extrapolate away from the wells.

Generally, however, the situation is the reverse of that just considered, and well data is sparse while stacking velocities are available along the seismic lines which are being interpreted. It is possible to perform depth conversion directly from the rms velocities to the various horizons obtained from the velocity analysis. Unfortunately, contours of rms velocity to a particular horizon are only loosely correlated with the structure on that horizon, owing to the variety of effects of depth of burial and changing formation thicknesses above the horizon. It is therefore

difficult to assess the geological plausibility of the details of the contour maps, and to pick out the anomalous velocity analyses (perhaps situated in a region of tectonic disturbance). Also, the rms velocity will be slightly different from the true average velocity to a particular horizon; the magnitude of the discrepancy depends on the velocity gradients present, and a correction may be required in accurate work. Therefore, it is generally better to work out interval velocities for the various formations and contour these, smoothing them severely if necessary to obtain geological plausibility. Usually the rms velocities are not corrected for dip (see p.44), because unmigrated sections tend to under-estimate the two-way-time to a dipping event, which is partially compensated for (exactly in the case of a plane layer) by the over-estimated velocity derived from velocity analyses not corrected for dip.

Problems arise if velocity data from more than one survey are to be used. While they should agree in general trends, they will usually not agree exactly on numerical values; *ad hoc* adjustment is called for, which can often be carried out by hand contouring of the two sets of data

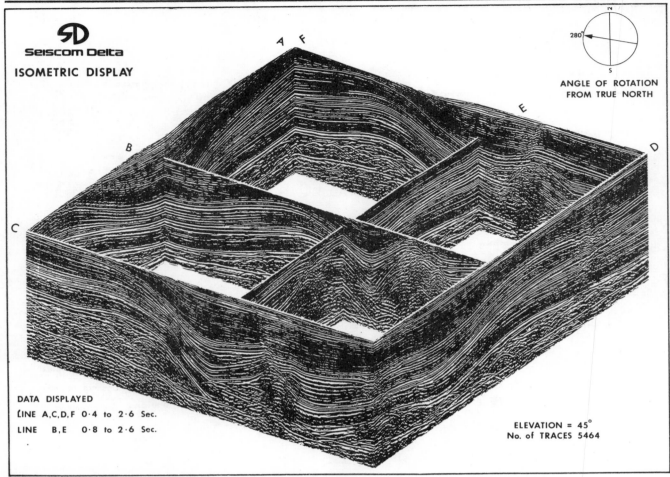

ISOMETRIC DISPLAY

DATA DISPLAYED
LINE A,C,D,F 0·4 to 2·6 Sec.
LINE   B,E   0·8 to 2·6 Sec.

ANGLE of ROTATION
FROM TRUE NORTH

280°

ELEVATION = 45°
No. of TRACES 5464

Figure 5/8: Isometric section projection of a group of seismic 'loops' prepared to aid three-dimensional comprehension of a structurally complex area.
(*Courtesy: Seiscom Delta*).

posted on the same map. Ties with well velocity data are also seldom satisfactory, and again *ad hoc* adjustment is called for. Where check-shooting has been carried out at a well (see p.61), the agreement will be better, but seldom perfect; the discrepancies remaining are partly a consequence of the difference in geometry, namely, normal incidence for the check-shot and various angles of incidence for the CDP gather. Therefore, horizontal variations in interval velocity and also anisotropy (commonly horizontal velocities are a few per cent higher than vertical) will cause small differences between well and seismic velocity estimates.

If no other reliable information is available, it is sometimes necessary to use an average velocity for a particular formation, deduced from a knowledge of its lithology.

5.7 Depth conversion

Given maps of two-way-time to various horizons and of the average velocities between the formations, it is in principle simple to multiply times by velocities to arrive at depth, building up the depth to each horizon in turn by a 'layer-cake' method. These depth values can then be posted and contoured in the same way as the two-way-time values.

In practice, the amount of computation required if a large area is to be mapped on several horizons is formidable,

and it is at this stage that a computer becomes extremely valuable. Various approaches can be used for the automated calculation of depth; the main problem is to make sure that the depth conversion is carried out in such a way as to ensure consistency of values at line ties and near wells. One method is to represent the velocity data by a series of values on a square grid, the grid size being small enough to represent adequately the complexity of the data. Such a grid can be constructed by reading off values by hand from the velocity contour maps, or can be machine-generated from the contour map after the contours have been digitised. At each shot point where depth conversion is required, the computer than calculates an average velocity value by interpolation from several neighbouring grid points, which is used in the depth calculation for that shot point. An advantage of this method is that well data can be included together with the grid points, and can be heavily weighted so that lines in the vicinity of wells will tie into them exactly.

5.8 Isopachs

Isopachs may readily be constructed by subtracting the time or depth values to two different horizons at each shot point. They are often useful in assessing the geological history of an area particularly in unravelling the history of sedimentary deposition. Again, the work involved in manual computation of isopach values can be considerable, and such maps are more readily generated by the computer from the digitised seismic information. For a fully automatic system, however, it is necessary to have a rather sophisticated system of reflector nomenclature if isopachs are to be correctly calculated in the case where outcrop or faulting makes for the disappearance of some formations over part of the map area; this implies that a complex naming system must be used in the header information of each seismic line, increasing the possibility of digitiser operator error.

5.9 Reporting and management presentations

The primary product of a geophysical interpretation of a set of seismic sections is a series of maps; two-way-time maps, depth contoured maps and, in some cases, isopach maps. In oil prospecting, these maps will be annotated to indicate the presence of prospective highs, or other potential trap structures. In mining and engineering investigations, important features such as faults, steep dips, structural axes, pinch-outs or interval thickness variation are likely to be the important features which need highlighting on the finally drafted maps. In either case it will be necessary to present these maps, along with an interpretation report to either the client who has commissioned the interpretation, or to company management, with recommendations for further investigation or development of the prospect under investigation. At this stage close collaboration between geologist and geophysicist is essential.

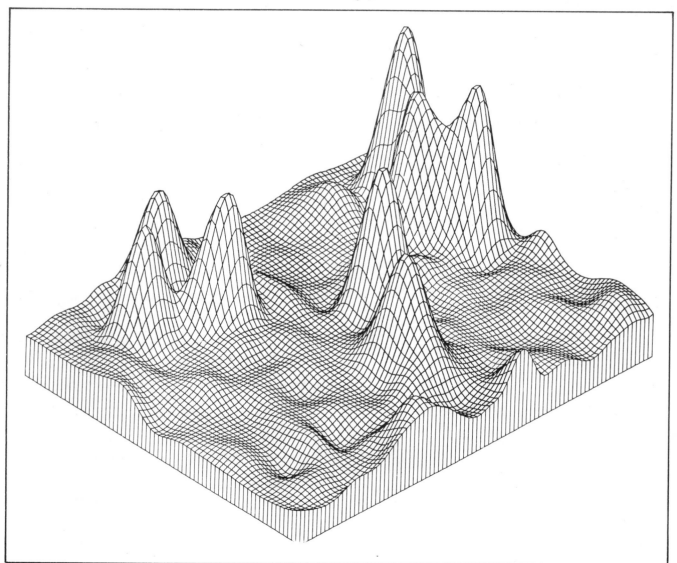

Figure 5/9: Isometric projection of a Devonian reef surface in Western Canada. Vertical scale is greatly exaggerated, the maximum relief being approximately 180m; the grid square size is approximately 100m.
(*Courtesy: Mobil Oil Canada Ltd*).

A typical report will include most of the following sections:

1. An introduction describing aims of the interpretation.

2. A description of the data used with notes on acquisition and processing parameters if relevant, and comments on the quality of the data. If static adjustments have been necessary to allow fit of data from different surveys, these should be noted.

3. A description of the geology and structural framework of the area, illustrated with relevant diagrams.

4. A section describing ties between wells and the seismic sections including a list for each well of seismic times to the tops of important horizons; in particular those which have been mapped.

5. A section on the selection of marker horizons which have been mapped giving evidence for stratigraphic identification of these horizons. It is usual to present sample interpreted sections to illustrate typical structural features and reflector quality. Horizons may have been selected because of prospectivity significance rather than reflector quality.

6. A detailed description of the interpretation of each horizon mapped — this usually describes the main features of the map or set of maps associated with the horizon in question.

7. A description of the main prospective structures (in oil exploration) or mining or engineering hazards etc depending on the aim of the investigation.

8. Recommendations for further work; more seismics, other geophysical investigations, drilling. If drilling is recommended a prognosis will be necessary for each suggested site giving a best estimate of the expected geological section. Estimates of depths are necessary to those intervals where cores will be required, as well as to those of principal economic interest.

9. The aim of the study may be to provide a more general prospectivity assessment of the area to give guidance to management of future development policy, in which case a classification and grading of structures will be required. Detailed drilling recommendations would then depend on follow-up surveys and these would only be undertaken if the company concerned was able to acquire the necessary exploitation licence.

10. A summary of work undertaken, lists of maps and technical appendices.

Other sections of a specifically geological nature may be necessary to supplement the geophysical interpreter's input to such a report. Visual aids are extremely important in conveying an interpreter's concepts and based on the principle that a picture is worth a thousand words, any illustrations that reduce the complexity of the seismic data to a simple, understandable form should be included; in particular, any specialised data processing such as described in chapter 3 or successful modelling as described in chapter 8 should be highlighted. The three-dimensional aspects of an interpretation, often difficult to grasp by the casual or busy observer, can be enhanced by the use of

isometric views of key horizons as seen in figure 5/9.

A seismic interpretation report, particularly of a large offshore area, can be quite voluminous. Presentation of the main conclusions of a report to management, partners, clients or government licensors etc can be accomplished in a visually effective manner by the preparation of a montage. Using, for example, a one metre high reproducible base, designed to go through a normal dyeline machine, and other supporting geophysical information (such as gravity or magnetics) or seismic modelling can all be brought together.

References and suggested reading

A.A. Fitch, *Seismic reflection interpretation.* (Gebruder Borntraeger, Berlin, 1976).

A. Garotta, 'Selection of seismic picking based upon the dip, move-out and amplitude of each event' *Geophys. Prospect.,* 19 (1971), pp. 357-70.

W.H. Hintze, 'Depiction of faults on stratigraphic isopach maps' *Bull. AAPG.,* 55, (1971) p.871.

J.T. Hornabrook, 'Seismic re-interpretation clarifies' *Petrol International,* 14 (1974) pp. 45-53.

D. Paturet, 'Different methods of time-depth conversion with and without migration' *Geophys. Prospect.,* 19, (1971), pp. 27-41.

K.V. Paulson and S.C. Merdler, 'Automatic seismic reflection picking' *Geophysics,* 33, (1968), pp. 431-49.

Prakla-Seismos, *Migration of reflection-time maps.* Prakla-Seismos Information No. 1, (Prakla Seismos, Hannover, 1977)

Seismic interpretation (Prakla Seismos Hannover).

W.M. Telford, L.P. Geldart, R.E. Sheriff and D.A. Keys, *Applied Geophysics.* (Cambridge University Press, Cambridge, 1976).

P. Tucker and H. Yorston, *Pitfalls in seismic interpretation.* SEG Monograph No. 2. (1973).

6. GEOLOGICAL INTERPRETATION

Interpreting seismic sections, producing time, depth and isopach maps is a task which depends on the interpreter's ability to pick and follow reflecting horizons (reflectors) across an area of study. He has to be able to correlate across faults and across zones where reflectors are absent because of geological discontinuity, for example, between basins either side of a structural high or between fault blocks beneath an unconformity. Reflectors usually correspond with horizons marking the boundary between rocks of markedly different lithology. Such a boundary does not always occur exactly at a geological horizon of major chronostratigraphic importance, such as the base or top of a System or Series, but may be simply a seismic marker horizon which occurs close to that boundary. This problem can be resolved by correlation of seismic and borehole data; synthetic seismograms have particular relevance to such studies. In this chapter we shall discuss the elements of structural geology and lithological variation in sedimentary rocks as discernible on seismic sections and in those respects which are important to the problems of seismic interpretation.

6.1 Lithology of common sedimentry rocks

As stated above, we are principally concerned in seismic interpretation with the structure in thick accumulations of sedimentary rocks. However, firstly we will consider one term in common use in seismic interpretation, that of economic (or seismic) basement. The term basement is also widely used in geology to signify a complex of crystalline and metamorphic rocks which are covered unconformably by unmetamorphosed sediments. Often the two terms are synonymous within a particular area of study, whilst in other circumstances the seismic horizon which is mapped and termed as basement may not in fact be the base of the sedimentary succession, it may, for example, be the top of a series of lavas or sills, intruded into the sediment, the top of which is the lowest horizon which can be effectively mapped by the seismic method. In some circumstances, this may, in practical terms, form the economic basement as far as exploration for hydrocarbon reservoirs is concerned. On the other hand, it is not unusual for igneous rocks, such as volcanic sills, lavas and tuffs, to occur as important markers within a stratigraphic sequence in circumstances where lower horizons and structures are easily identified; in this case the top of the igneous rock layer would not be termed the basement horizon. Care must therefore be taken in interpreting the use of the term basement both in well logs and on seismic maps.

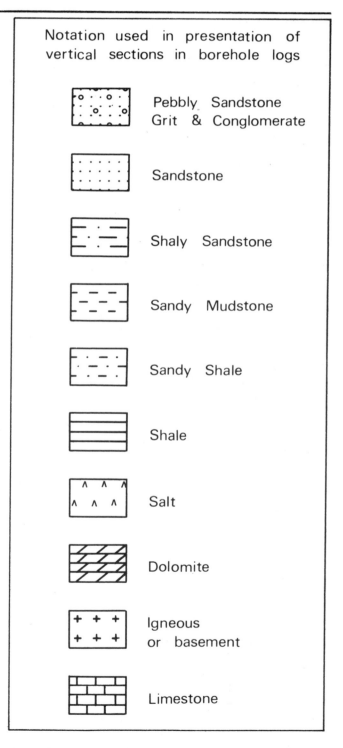

Notation used in presentation of vertical sections in borehole logs

Pebbly Sandstone Grit & Conglomerate

Sandstone

Shaly Sandstone

Sandy Mudstone

Sandy Shale

Shale

Salt

Dolomite

Igneous or basement

Limestone

Figure 6/1: Typical lithological notation.

Thick accumulations of sedimentary rocks occur in large scale geological structures termed sedimentry basins (see 6.2) and the material which forms the major component of infill of such sedimentary basins is a group of rocks known as the clastic rocks. These rocks are formed as a result of weathering of pre-existing rocks into fragments of varying size and composition which are then transported from the area of origin to the area of deposition. Clastic rocks are thus built up of fragments and are classified according to the size and mineralogy of these grains. A commonly used particle size classification is the Wentworth Classification (see table 6/1). Most sedimentary rocks are a mixture of grains of various sizes, the bulk of the rock may, for example, be made up of sand grains but infilling spaces between these grains may be a matrix of much finer rock fragments and finally the rock may have been cemented by a material, usually of chemical or biochemical nature which has impregnated the rock and bonded the grains together during the process of diagenesis. In figure 6/1 we see a typical notation used in preparation of composite logs to indicate lithological variation within a borehole section; this diagram shows only a selection of the many rock types which may occur. The uppermost six rock types are varieties of clastic rock, the coarsest being the pebbly sandstones, grits and conglomerates (breccias are similar to conglomerates but with angular as opposed to rounded rock fragments). Sandstones are common reservoir rocks,

especially if relatively uncemented and clean, that is, they do not contain a high proportion of fine-grained matrix minerals. Sandstones are of the arenaceous group of rocks in which grain sizes range from 1/16mm to 2mm, and can be graded from very fine to very coarse. Sandstones have grains predominantly of quartz but other arenaceous rocks can be important, such as arkose rocks which contain notable quantities of feldspar grains in addition to quartz, also greywackes which are poorly sorted rocks made up of a wide range of grain material. Fine-grained sedimentary rock types are usually termed argillaceous; such rocks include clay, shales, mudstones, siltstones and marls. They are usually porous, but, because of their fine grain, they are in most cases poorly permeable and often provide cap rock for hydrocarbon reservoirs. Shales are well laminated mudstones and siltstones are similar to mudstones but with a predominance of slightly coarser grade silty material. Marls are calcareous mudstones. Such terms as shaley sandstone shown on a borehole log usually indicate a rock with layers and partings of shaley and sandstone rock material, the bulk being a compound of both rock types.

Apart from clastic rocks, the other two main groups of sedimentary rocks to be considered are the chemical and organic deposits. Chemical deposits can be of great importance in the development of hydrocarbon-containing sedimentary basins. These deposits are formed by chemical precipitation from aqueous solutions. Organic deposits are

Table 6/1: Particle Size Classification (Wentworth).			
Classification	**Grade Limits (Diameters in mm)**	**Microns**	**Retained on Mesh**
Boulder	Above 256 mm		
Large cobble	256–128		
Small cobble	128–64		
Pebble			
Very large pebble	64–32		
Large pebble	32–16		
Medium pebble	16–8		
Small pebble	8–4		5
Granule	4–2		6
Sand			
Very coarse sand	2–1		12
Coarse sand	1–1/2		20
Medium sand	1/2–1/4		40
Fine sand	1/4–1/8		70
Very fine sand	1/8–1/16	125–62.5	140
Silt			
Coarse silt	1/16–1/32	62.5–31.2	270
Medium silt	1/32–1/64	31.2–15.6	
Fine silt	1/64–1/128	15.6–7.8	
Very fine silt	1/128–1/256	7.8–3.9	
Clay			
Coarse clay	1/256–1/512	3.9–1.95	
Medium clay	1/512–1/1,024	1.95–0.975	
Fine clay	1/1,024–1/2,048	0.975–0.487	
Very fine clay	1/2,048–1/4,096	1.487–0.243	

formed by deposition of fragments of material which were originally parts of living organisms, for example, some organic reef limestones contain many shell fragments, coral remains etc. Some common rocks form through contemporaneous accumulation of organic fragments coupled with chemical deposition from solutions. Carbonates are often this type and in petroleum geology these rocks along with sands and sandstones form most of the world's hydrocarbon reservoirs.

The carbonate group of rocks includes limestone (main constituent calcite: $CaCO_3$) and dolomite (main constituent being the mineral dolomite: $CaMg(CO_3)_2$) as well as intermediate rocks containing both calcite and dolomite minerals such as dolomitic limestones. Other rocks in the group contain varying proportions of such impurities as sand, silt and clay and are accordingly named; for example silty limestone. Chalk is a particular variety of limestone, usually very pure, with a high percentage calcite content. As reservoir rocks, the carbonates differ from sands and sandstones in that the rocks usually have relatively low porosity and permeability. Some porosity may be original as is the case in detrital limestones but more usually in carbonate reservoirs the necessary porosity and permeability occurs as a result of post-depositional solution and fracturing.

Other chemical deposits which can form important constituents of a sedimentary basin sequence are the group of chloride and sulphate rocks which form as a result of the evaporation of saline water. Such rocks are not important as reservoir rocks but can be important as cap-rock material by virtue of their low permeability. Rocks in this group include rock-salt (NaCl) sylvite (KCl) and carnallite ($KCl,MgCl_2.H_2O$) and these have additional importance in that under certain conditions they have the properties of a plastic, as opposed to an elastic, solid. Deformation and flow migration of these rocks can then bring about development of structures which form traps for hydrocarbon accumulation (see p.115). The sequence of rocks formed by evaporation of saline water is often termed the evaporite sequence. Typically in a sedimentary basin this might include layers of rock of the types: chemically deposited limestones, both calcitic and dolomitic, followed by gypsum ($CaSO_4.H_2O$) or anhydrite ($CaSO_4$) then, usually as a major component, rock-salt with, at the uppermost end of the depositional cycle, deposits of potash and magnesium salts such as sylvite and carnallite.

The last group of sedimentary rocks to be considered are the pyroclastic deposits. These are sediments which consist predominantly of volcanic ejectamenta. They can be deposited on land or in water. Lithology can range from coarse volcanic agglomerates to very fine deposits usually known as ashes or tuffs. These latter rocks are deposited from the clouds of very fine rock particles which are formed during explosive volcanic eruptions. Such clouds can distribute material over very wide areas and tuffs and ashes can act as valuable marker horizons in seismic interpretation as well as being an invaluable aid in the task of correlation between well logs in tectonically active areas. Such rocks are not generally important either as reservoir rocks or as cap material even though porous and permeable pyroclastic rocks do exist.

Most sedimentary basins are formed over very long periods of time, often with stages of both marine and non-marine deposition, and including sedimentation in both shallow water and deep water environments. These environmental changes are reflected in the type of rocks formed; cycles of sedimentation can be detected from which the history of the basin may be deduced. Many of the events which control changes in sedimentation type occur abruptly thus changes from one rock type to another are often seen in a rock succession at distinct horizons without gradation. This does not necessarily indicate any cessation of deposition between one bed and the next. With continued development of a sedimentary basin earlier sediments become more deeply buried. Sedimentation is followed by a long period of diagenesis. Diagenetic processes are those which affect the rock during burial and bring about its compaction and cementation. These are relatively low temperature, low pressure processes (as opposed to metamorphic processes) which convert the unconsolidated sediment into a mature rock. Following diagenesis, the mature rocks will have varying physical properties and in particular varying densities and elastic modulii. In general there is a tendency for seismic velocity — and to some extent density — to increase with depth of burial. These changes are not gradual however, and as we have noted previously (p.3) it is at the boundaries between units of rocks of significantly different velocity and density that strong seismic reflections occur. Depending on the history of the basin, a seismic reflector associated with change of lithology at a boundary between rock units can be very widespread or restricted to a limited area or areas.

6.2 Depositional features of sedimentary rocks

Only fairly large-scale features of the depositional structure of sedimentary rocks can be studied using reflection seismics. Internal structure within thin beds of rock are not generally discernible. Basins are formed by regional subsidence of an area and one of the results of a seismic interpretation such as described in the previous chapter is that it provides the geologist with a picture of large-scale variations in thickness of the main rock units. This gives evidence on the mechanism of subsidence (whether or not fault controlled); where maximum subsidence has occurred, as well as detailed evidence of the present-day shape of the basin. Basins can be either symmetrical or asymmetrical and both graben and half-graben structures are common, though faulting may post-date the main period of sediment accumulation (see figure 6/2). In studying the large-scale features of a basin it is often valuable to attempt to reconstruct the shape at various stages in its history and interpreted seismic profiles across the entire area, especially if controlled by well data, give exactly the information required to undertake such a reconstruction; see for example figure 6/3.

Seismic sections, as well as giving data on the macrostructure of a sedimentary basin can also give evidence on the environment at various stages of its development. This evidence relates in the main to non-tectonic structural features. Seismic profiles can indicate the presence of marine transgressions and regressions; evidence of shallow water sedimentation is seen in such features as cyclic sedimentation and the presence of infilled river channels, deltaic deposits and coastal features such as sand bars; evidence of deep water sedimentation is seen in the presence of thick uniform sections of homogeneous generally fine-grained rock in places associated with

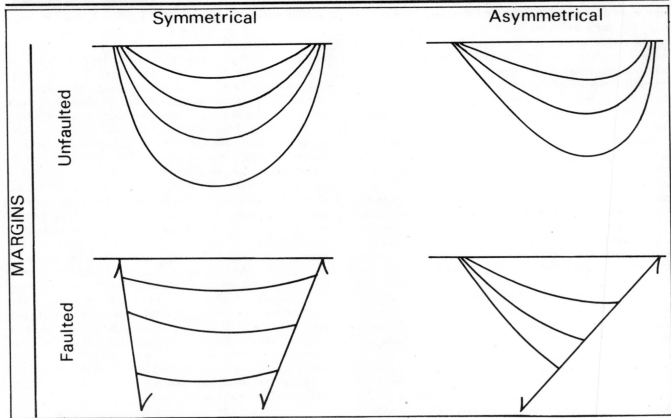

Figure 6/2: Classification of sedimentary basins (after Chapman).

marginal accumulations of turbidites. Evidence of halokinesis (see p.89) indicates firstly a continental depositional environment during formation of evaporites in which the basin had only restricted connection with ocean waters. Secondly halokinesis indicates the eventual burial of the evaporites to a depth sufficient to promote mobilisation of the sodium and potassium salt formations. Ancient desert sands can be identified through seismic structure attributable to dune bedding. Also, the presence of reefs, either of barrier type or reef-knolls, indicates a particular type of shallow marine water environment at the time of deposition. Examples of seismic sections showing a number of the sedimentary features noted above are illustrated in the case studies in chapters 9—12, see for example the reef structure in figure 10/7b and the complex system of channels on the section in figure 11/16. The object of the above discussion is to stress the importance of co-operation between regional geologist and seismic interpreter in viewing of the stratigraphy and structure and historical development of any basin under study. This is particularly true where limited borehole data are available and any regional geological interpretation must depend on seismic evidence as an indicator of the variation in sedimentary environment. Depositional features can be recognised on seismic sections which are not evident from any study of interpreted seismic maps.

6.3 Deformation mechanisms
As well as giving an indication of the depositional history of sedimentary rocks formed within basins, seismic sections also give clear evidence of any major deformation which has affected such rocks. The forces which cause deformation can be both compressional and tensional and have both vertical and horizontal components. In some circumstances gravity is an important factor in producing deformational stresses. Rocks may be deformed to produce both small and large-scale structures; in seismic interpretation we are concerned only with the large-scale structures which can be broadly classified as being either faulting or folding. It is not uncommon however for structures to develop which combine both types; a deep seated fault for example may be draped at a higher level by a monoclinal fold. In general, rocks are more likely to fracture with the consequent development of faulting when subject to tensional or shear stresses than when subject to a primarily compressional stress, though thrust faulting in particular is usually closely associated with compression and folding in stratified rocks.

6.3.1 Faults
The term fault describes the displacement of a body of rocks by shearing or fracturing along a planar surface which is called the fault plane. In many situations the fault plane is not a simple surface but a zone of crushed rock which may range from a few centimetres to hundreds of metres wide depending on the magnitude and type of fault involved. On seismic sections, faults are identified where reflectors can be seen to be displaced vertically, or (less usually) by identification of zones of crushed rock. The interpreted sections in figure 9/7 show numerous examples of fault structure.

Geological interpretation of seismic data, not only sections but also isochron and depth maps, requires description of faulting in terms of standard geological nomen-

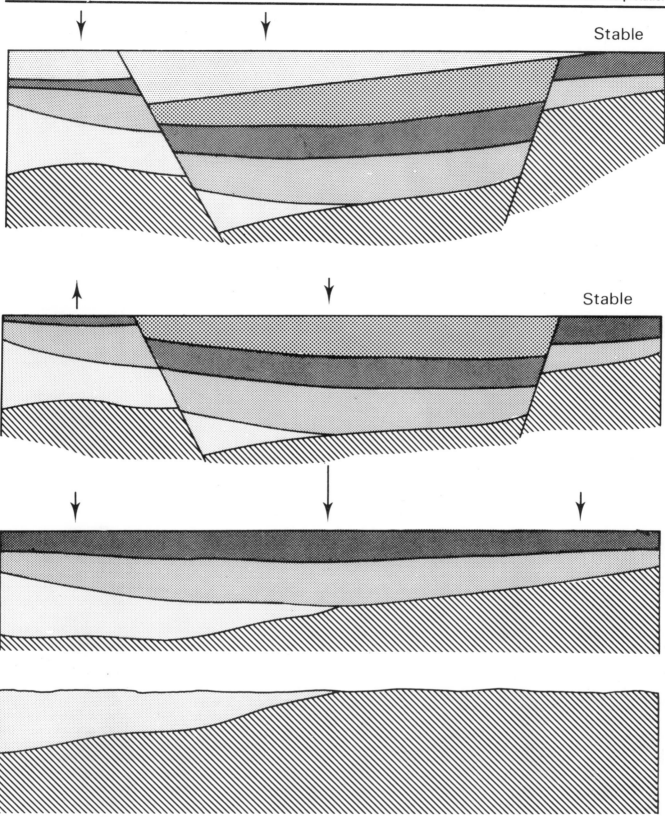

Figure 6/3: Reconstruction of the history of a sedimentary basin from seismic data. The schematic geological section at the top represents a seismic interpretation of present-day configuration of rock units. This structure is estimated to have developed from the earliest stage, the bottom section, through stages of vertical movement (indicated by arrows), sediment infill and erosion.

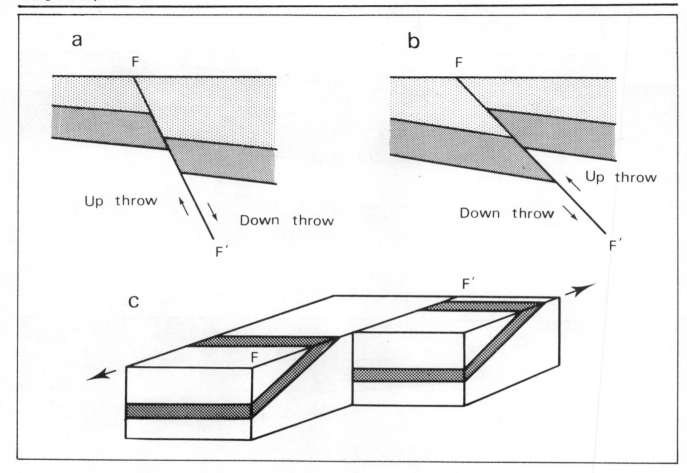

Figure 6/4: Classification of types of faults, **a)** normal fault, **b)** reverse fault, **c)** strike slip fault. Arrows show direction of movement. F—F' is the fault plane.

clature. Faults are usually classified as being of the types normal, reverse (thrust) or strike-slip. Figure 6/4 shows diagramatically the types of displacement involved. Normal faults are generally associated with tensional stress, reverse faults with compressional stress and strike-slip faults with shear stress. The directional trend of a fault is termed its strike and the angle of inclination of the fault-plane with respect to horizontal, its dip..

Alternatively, the hade of a fault may be quoted, this being the fault-plane angle with respect to the vertical (see figure 6/5). The amount of vertical displacement associated with a fault at any location is termed the throw of the fault. Strike-slip faults are often termed alternatively as wrench faults, tear faults or transcurrent faults and displacement of such a fault, if it can be assessed, is quoted as a horizontal displacement which in some cases may amount to tens or even hundreds of kilometres. The direction of displacement is designated as either dextral or sinistral, a dextral fault being one in which displacement of the far side of the fault is to the right as viewed by an observer facing the fault, and a sinistral fault is one in which this displacement is to the left (figure 6/6).

During the process of development of a sedimentary basin, periods of faulting are more likely to be episodic than continuous. Nevertheless, the fact that vertical displacements of major rock units are likely to occur con-

temporaneously with sedimentation is of fundamental importance to the history of any basin development. Fault blocks which are structurally elevated, and such structures can be identified as such on seismic sections, are termed horst blocks whereas faulted depressions are termed grabens. An asymmetrically developed graben is often termed a half-graben (see figure 6/7). When development of any fault system is contemporaneous with sedimentation within a basin, the faulting is associated with growth structure and such structure has particular relevance to hydrocarbon accumulation.* For example, growth faults are often associated with antithetic faulting (see figure 6/8 and the seismic section in figure 7/24). The combination of normal faulting, antithetic faulting, growth structure and 'roll-over'(that is the development of an anticlinal trapping mechanism in association with a growth fault) can provide all the principal requirements for development of a hydrocarbon pool. Growth structure over horst blocks and in association with normal faulting can also provide the mechanism for both hydrocarbon fluid migration and formation of traps. In oil exploration, the geological interpretation of seismic data is therefore not just a process of defining potential structural traps but also that of inter-

---

*See R.E. Chapman *Petroleum Geology. A concise guide* (Elsevier, Amsterdam, 1973).

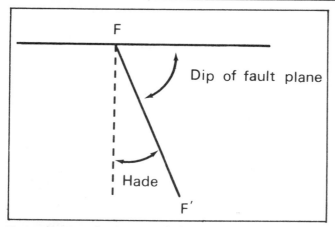

Figure 6/5: Dip and hade angles of a fault.

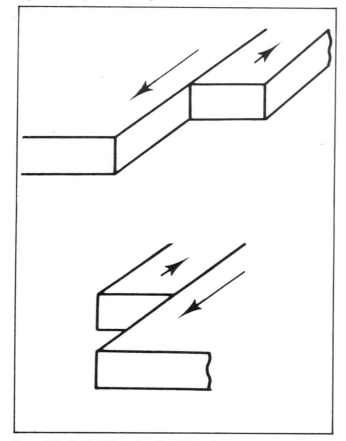

Figure 6/6: Strike-slip faults, **a)** sinistral, **b)** dextral.

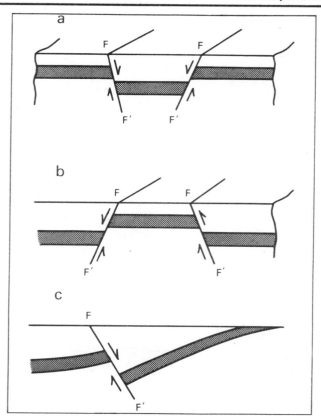

Figure 6/7: **a)** Graben, **b)** Horst, **c)** Half-graben.

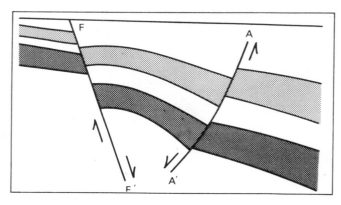

Figure 6/8: Normal growth fault F—F' with antithetic fault A—A'.

preting the structural development of a basin or province so that its history may be related to the many factors which contribute to its hydrocarbon prospectivity.

6.3.3 Unconformities

When deformation occurs within a sedimentary basin contemporaneously with the long term process of its infilling by sediments, it is likely that in at least parts of the basin the process of deposition of sediments will be at times interrupted. Such interruptions can be recognised as periods of non-deposition or removal of previously deposited material by erosion. When such a period is followed by further deposition of sediments, the surface

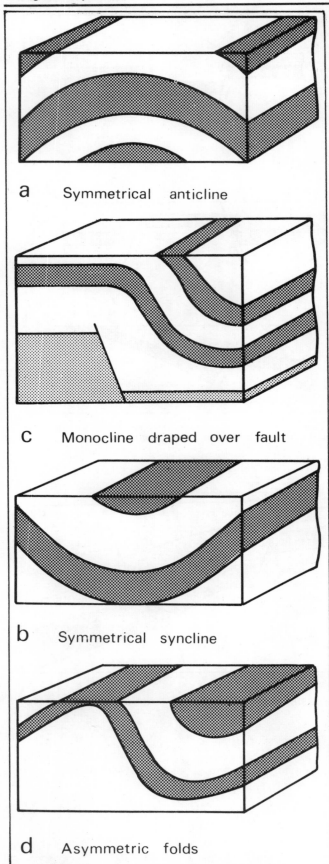

a Symmetrical anticline

c Monocline draped over fault

b Symmetrical syncline

d Asymmetric folds

Figure 6/9: Types of folds. **a)** Symmetrical anticline, **b)** Symmetrical syncline, **c)** Monocline draped over fault, **d)** Asymmetric folds.

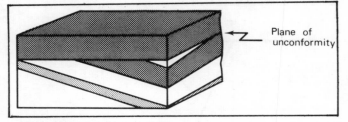

Plane of unconformity

Figure 6/10: An angular unconformity.

which marks the break in deposition is termed an unconformity. Unconformities can be caused by processes other than deformation of rocks within a basin, such as major climatic changes, changes of sea-level or tectonic activity outside the basin, in which case unconformities are recognised which, for example, can be seen as periods of basin-wide non-deposition.

On seismic sections, unconformities can be recognised easily only when the strata beneath the plane of unconformity lies at an angle to those above it, see figure 6/10. Note also the development of an unconformity as illustrated in figure 6/3 and the unconformity which can be recognised in the seismic section of figure 7/3.

6.3.4 Diapirism and salt tectonics

Some sedimentary materials, primarily salt and clay, have the property that, under certain conditions, bodies of such rocks will deform by plastic flow, and migration can take place both vertically and horizontally. Diapirs are said to have formed when this process leads to intrusion of the migrating plastic sediment body upwards through overlying strata to a level of equilibrium higher in the rock succession, or in some cases by extrusion to the earth's surface. Rock flow structures are most easily recognised on seismic sections where horizons can be mapped at levels below the rock units which have been deformed by migration. The structure of underlying formations will be seen to be relatively undisturbed and largely independent of the structure in rock units overlying the deformed clay or salt. In the case of salt tectonics, strata overlying the salt may be deformed into anticlines, synclines, domes, faults and overthrusts, whereas beneath the salt layer the beds can be undisturbed and flat-lying. A typical salt diapir structure is illustrated in figure 8/4. Clay diapirs are less common, and clay migration usually takes the form of flow into anticlines as a result of folding associated with compressive tectonic forces.

6.3.2 Folds

Whereas the interpretation of fault patterns, as seen on seismic sections, and the synthesis of interpreted sections into seismic horizon maps can often pose many problems, particularly those of correlation across fault lines, the identification and mapping of folds, dome structures, anticlines, synclines and monoclines is usually a relatively straightforward interpretational task. Problems can arise however at the stage of compiling isochron maps, particularly where structure is complex and both folding and faulting have affected the rock strata. As described in chapter 5, it is often useful before commencing a detailed interpretation of a set of seismic data, to prepare a map of principal structural features and trends (see figure 5/2) and such a map can be particularly helpful in establishing as early as possible, the trends of fold structures.

Any flexure of rock strata might be termed a fold. Only large-scale folds can be detected and mapped by the seismic method. Various types of fold are shown in figure 6/9. It should be noted that the simple classification of folds into the three groups, anticlinal, synclinal and monoclinal, would need to be expanded if more deformed structures were being described than can normally be seen on a seismic section. Many folds are asymmetric, thus when horizon maps of different levels are being compared, the axes of anticlines and synclines will be seen to have different positions. When the axis of a fold is not horizontal, it is said to plunge in a particular direction, and the angle between horizontal and axis is termed the angle of plunge. This angle will vary along the axis of a fold and may also vary between horizon levels in the fold due to variations in bed thickness along the fold axis. Doming of strata, as is often caused by salt movement through formation of diapirs, is a type of anticlinal development where there is no well defined fold axis.

As with faulting, deformation by folding can occur at various times during the history of deposition of a sedimentary basin; in which case growth structures are likely to be apparent with thinning of rock units over anticlinal axes and thickening within synclinal axes. Recognition of such growth structure helps in dating the periods of deformation. In hydrocarbon exploration such growth structures are important in that they are likely to correlate with lateral variations in rock type and with variations in rock properties such as porosity and permeability.

Compressive tectonic forces are not an essential requirement to the mechanism of salt migration. According to Trusheim,* many salt structures in northern Germany can be attributed to the autonomous movement of salt under the influence of gravity. These structures develop as a result of the comparatively low density of salt and its ability to flow as a result of overburden pressure alone. Such structures are termed halokinetic structures, whereas those which arise as a result of compressive tectonic forces are termed halotectonic.

Halokinetic structure can only develop where salt layers have been buried to such a depth that loading causes the salt to deform as a plastic (as opposed to an elastic) solid and experience in northern Germany indicates that an overburden of approximately 1000m thickness is necessary. Figure 6/11 shows the typical development of a salt stock. According to this reconstruction, salt movement has not only caused deformation of the overlying strata but has also strongly influenced the deposition of sediments during periods of structural growth. Both faulting and folding are associated with the deformation.

Salt structures occur in a wide variety of forms: stocks, domes, plugs, pillows and walls. Each structure may require detailed seismic mapping before its shape can be properly defined and in some circumstances specialised survey techniques are required to acquire good reflection data from below salt structures. Salt structures are important in oil exploration in that a wide variety of trap structures can develop in a region as a result of deformation by salt tectonics (see chapter 8).

6.4 Outcrop geology ties

In chapter 4 we discussed the importance of borehole data in establishing the geological significance of a seismic interpretation and how ties are made between seismic sections and well logs. Well data afford the most reliable means of providing geological control to a seismic interpretation. In situations where well data are not available or inadequate, it is necessary to attempt a geological interpretation of the seismic sections, including chronostratigraphic identification of principal reflectors, using such other sources of geological data as are available. One method is that of comparing the seismic sections to be interpreted with sections from other adjacent areas where good stratigraphic ties have been made with well data. This technique involves correlation of major unconformities, making comparisons of stratigraphy as indicated by seismic character, and identifying and correlating such features as marine regressions and transgressions. To be reliable, this demands a good knowledge of the regional geology around the prospect, and in particular of the stratigraphy of the area.

Another means of obtaining geological control is through tying the seismic interpretation to the outcrop geology of an area. The efficacy of such a tie will depend on how well the outcrop geology is known and whether structure at outcrop level can be reliably extrapolated downwards to allow prediction of structure at depth. There may well be difficulties associated with poor seismic data.

On land, outcrop geology is likely to be well established and the costs of revision mapping and additional shallow drilling are relatively low, so that outcrop geological mapping, or a study of existing maps and sections is generally accepted as an essential part of any exploration programme. At sea, the outcrop geology is usually less well known, and the costs of additional sampling and shallow drilling are high when compared with landward investigations. Exploration programmes at sea therefore do not often include any extensive survey of outcrop geology. Nevertheless, in both situations, it may be possible to add significantly to the geological validity of a seismic interpretation by careful study of, and tying to, known outcrop geology. An example of the value of establishing an interpretational tie to outcrop geology is described in chapter 9 where, as a consequence of the study, it was furthermore found necessary to update and modify the existing geological map.

One notable difficulty in tying reflection seismic data to outcrop geology is that in the very zone where correlation is being attempted, immediately adjacent to land surface or seabed, seismic data give poor structural resolution. In the vast majority of seismic surveys, both acquisition and processing parameters are generally defined so as to investigate structure down to greatest possible depths, or at least to the level of economic basement which may be at a few thousand metres below datum. Penetration to such deep horizons is often achieved at the expense of high resolution, particularly in the 0—500msec two-way reflection time zone. However, in some situations good ties to outcrop geology can be of equal benefit and less costly to obtain than ties which could be alternatively established by

---

*F. Trusheim, 'Mechanism of salt migration in Northern Germany' *Bull AAPG* 44 (1960) pp. 1519-1540.

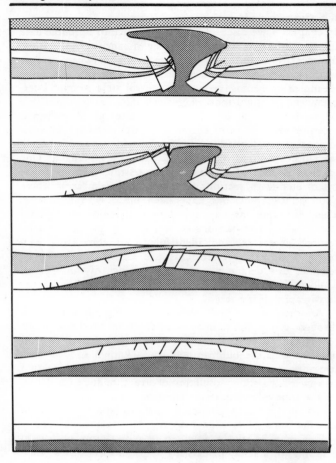

References and suggested reading

R.E. Chapman, *Petroleum geology. A concise study.* (Elsevier, Amsterdam, 1973).

A.I. Levorsen, *Geology of petroleum.* (W.H. Freeman & San Francisco, 1954).

W.L. Russell, *Principles of petroleum geology.* (McGraw-Hill, New York, 1960).

E. Sherbon Hills, *Elements of structural geology.* (Chapman and Hall, London, 1972).

F. Trusheim, 'Mechanism of salt migration in northern Germany' *Bull. AAPG,* 44 (1960), pp. 1519-1540.

D.G.A. Whitton and J.R.V. Brooks, *A dictionary of geology.* (Penguin, Harmondsworth, 1972).

Figure 6/11: Diagramatic development of an asymmetric Zechstein salt stock (after Trusheim, 1960).

drilling shallow stratigraphic test wells. Seismic surveys can in such circumstances be designed to give optimum data quality in the 0—1sec two-way reflection time interval. Design of high resolution surveys is described in chapter 2. At sea, the results of single-channel seismic profiling can provide a valuable aid to the establishment of outcrop ties, see chapter 7 and in particular, figures 7/2 and 7/3.

7. OTHER GEOPHYSICAL METHODS

As is stated in introduction to chapter 5, the interpretation of conventional reflection seismic records has, as its first object, the provision of a series of structural contour maps which describe the subterranean topography of selected reflecting horizons within the geological strata of the area under investigation. If, for this same area, the interpreter has access to data from other types of geophysical surveys, he can use these both as a direct aid to the task of interpreting his seismic sections, and as a means of improving, or extending, his geological interpretation of the conventional seismic results.

Although, in principle, all types of geophysical data might have such an application, here we shall consider only those commonly used: shallow penetration high resolution reflection seismic profiling, seismic refraction surveys, magnetic and gravity surveys. These same methods are often used in reconnaissance of regions previously unexplored by reflection seismics as a means of, and at relatively low cost, making a first evaluation of regional geological structure and economic prospectivity. Results can then be used to guide the planning of subsequent conventional seismic reflection exploration. Also, in recent years, it has become fairly common in offshore exploration, for seismic survey vessels to be fitted with magnetometers and gravity meters. If gravity and magnetic data are acquired concurrently with seismic reflection data, this can be achieved at relatively little extra cost additional to that of the seismic survey. Over both land and sea, regional magnetic surveys are usually made from aircraft, and profiles along seismic lines are best interpreted in conjunction with the results of such regional aeromagnetic surveys. Similarly, single gravity profiles along seismic sections are of limited value unless interpreted against the background of a high quality regional survey based on either a regular grid of lines (at sea) or a regular grid of stations (on land). Gravity and magnetic surveys should cover not just the immediate prospect area; local anomalies should always be interpreted against a knowledge of broad scale regional variations.

7.1 Shallow seismic profiling

High resolution, shallow penetration, reflection seismic surveys are of particular value as an aid to interpretation in offshore areas which are not blanketed by thick young superficial sediments and which have been surveyed by coring and shallow drilling to provide data on the stratigraphy of rocks cropping out at the seabed or sub-cropping close to it. Techniques have been developed which use relatively low cost equipment and which do not require computer processing of digital records; these techniques nevertheless produce well resolved seismic sections in the uppermost few hundred metres of strata beneath seabed.* Comparing the method with conventional seismic reflection

acquisition systems, the main differences are:

1. The seismic source is of relatively low power but is designed to have a broad spectrum with the main energy in a frequency band ranging from a few hundred hertz to a few kilohertz; such devices as the sparker, boomer, pinger, air gun and water gun sources are commonly used. These devices are operated at firing rates of usually between one half second and two seconds depending on source energy and penetration required. With a ship travelling at four knots this corresponds to a pop interval (and thus sub-surface coverage) of between approximately 1.5 and 6m.

2. Single channel streamers are commonly used, either containing a single hydrophone (for highest resolution) or a short linear group of hydrophones.

The main components of a typical system are shown in figure 7/1. A seismic section is produced in real-time aboard ship on a graphic recorder with analogue tape-recording as a back-up and to allow subsequent replay using altered signal processing parameters.

Seismic profiles so produced offer very limited penetration (usually only a few hundred metres) but the sections provide a very well resolved indication of structure in the zone immediately beneath seabed and it is in this zone that conventional seismic sections are usually very obscure. In chapter 2 we discussed how to improve resolution close to seabed using a conventional digital acquisition system adjusted in terms of source frequency, streamer tow-depth and digital sampling rate. Such specialised acquisition and processing is very costly when compared with single-channel profiling and would only be used if a combination of this with conventional acquisition proved inadequate. In figures 7/2, 7/3 and 7/4 comparisons are shown between conventional sections, high resolution multi-channel sections and sparker profiles, all from the English Channel. These comparisons illustrate how the sparker profile can be used to trace to exposure at seabed horizons which are well developed at depth in conventional sections which give only an obscure indication of seabed outcrop. This link between seabed and deeper structure is also well resolved in the high resolution seismic sections in figures 7/3 and 7/4, but these sections were acquired at much greater cost than the sparker sections. The drawbacks of poor penetration and the interference caused by multiples in the sparker section are also well illustrated.

---

* For a fuller description of this method see R. McQuillin and D.A. Ardus, *Exploring the geology of shelf seas* chapter 4 (Graham and Trotman, London, 1977).

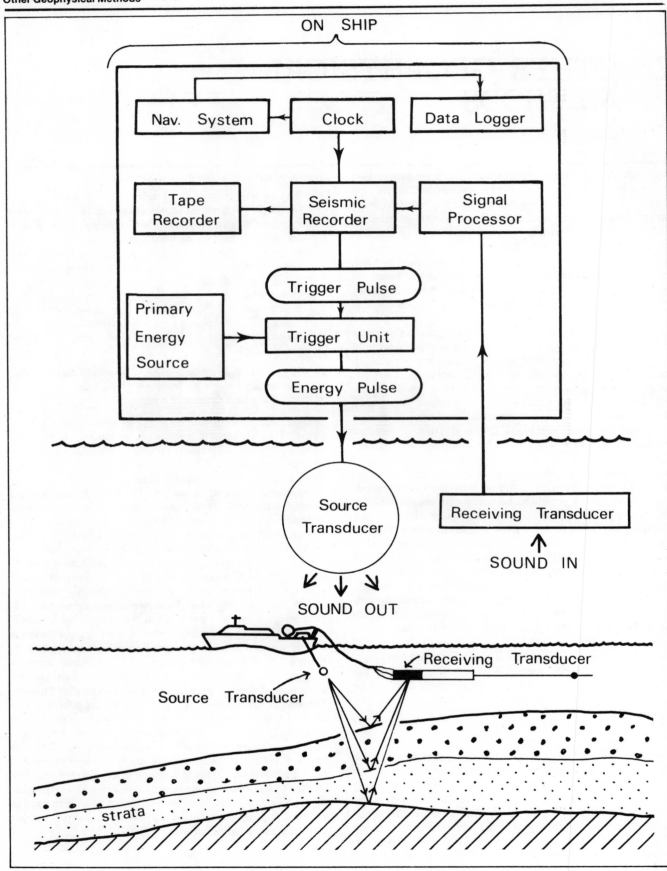

Figure 7/1: Block diagram showing the main components of a continuous sub-bottom profiling system.

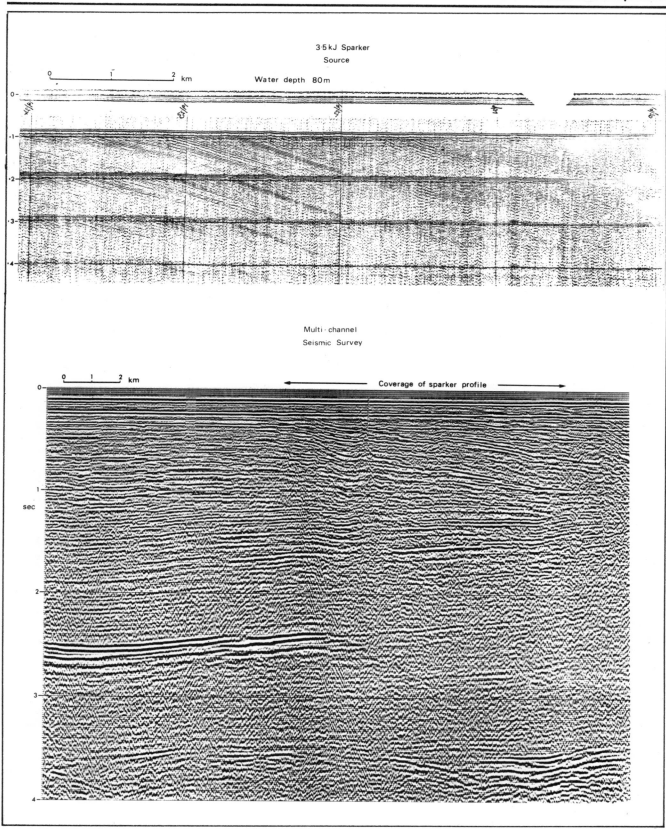

Figure 7/2: English Channel seismic reflection profiles; comparison of results obtained using a 3.5kJ sparker single channel profiling system with a conventional seismic record along the same line. Multi-channel section obtained using gas-guns as source, 2400m 48-channel cable and 24-fold processing.
(*Courtesy: IGS; surveys by IGS and Seiscom*).

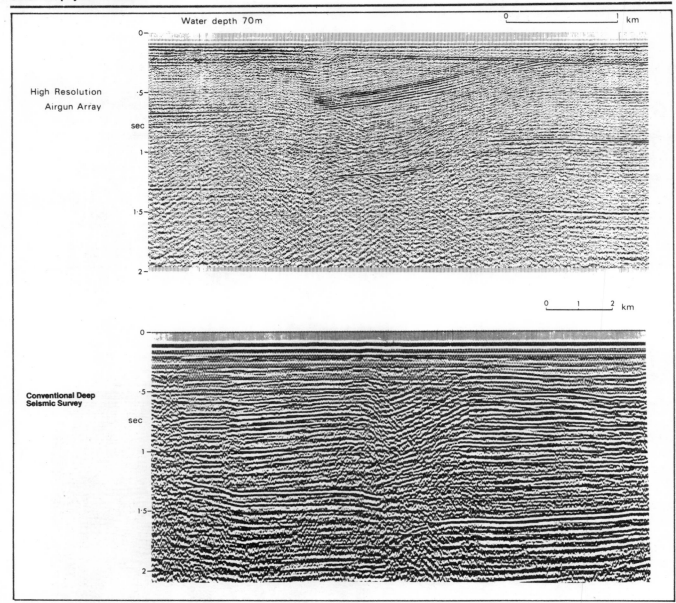

Figure 7/3: English Channel seismic reflection profiles; comparison of results of high resolution acquisition and processing with conventional seismic record. Conventional record from same survey as in figure 7/2, high resolution survey uses 600m 24-channel cable, 2ms sampling and processing to optimise retention of high frequencies.
(*Courtesy IGS; surveys by GSI and Seiscom*).

The area from which this example is taken is one where seabed geology is fairly well established by coring and shallow-drilling, results which have been interpreted in conjunction with those from a grid of sparker profiles to produce geological outcrop maps. These maps show major structures and stratigraphic boundaries. This known surface geology can then be used to provide stratigraphic identification of important reflectors seen at depth on the conventional seismic sections, a task which would have been more problematic in the absence of profiler data. Further, a study of the sparker sections gives information on the occurrence or absence of minor faulting and other small scale structures which cannot be resolved on the conventional sections but which might nevertheless be of sufficient magnitude to have an important bearing on the hydrocarbon reservoir characteristics of the area.

To summarise, profiling data can be used to clarify the interpretation of conventional reflection data close to seabed, they can be used to illustrate small scale structure, too fine to be resolved by the conventional method, and they can be used to provide information on the geology of the seabed, subcrops beneath superficial sediments and the subcrop geology beneath shallow unconformities. The data are also of value in planning well sites and have a wide range of engineering applications offshore.

7.2 Refraction seismics
In the early days of seismic exploration, refraction seismics

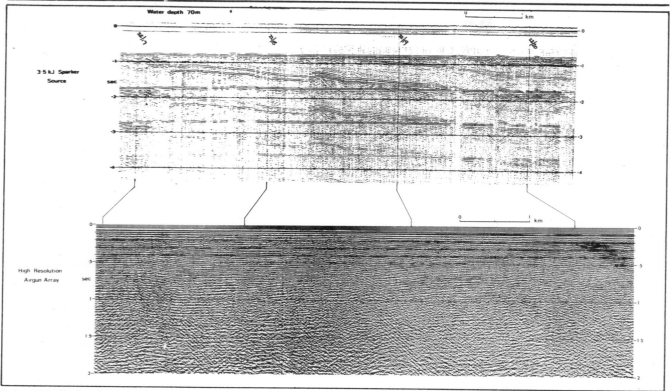

Figure 7/4: English Channel seismic reflection profiles; comparison of single channel sparker and high resolution multi-channel seismic sections.
(*Courtesy IGS; surveys by IGS and GSI*).

provided an important contribution to a high percentage of hydrocarbon exploration programmes. Even as recently as the late 1950s and early 1960s the method was in vogue in many parts of the world, both on land and at sea. Improvements in the reflection technique have however demoted the refraction method to one of only occasional commercial use, though it is still widely used in studies of deep crustal structure and studies of regional tectonics, mainly as one of a range of methods employed in such studies by government or university research groups. Today, the seismic interpreter is seldom likely to be called upon to integrate the results of refraction surveys with an interpretation of reflection data. However, if refraction data are available in an area which is being subject to interpretation, these data can be a useful supplement to the results of reflection seismics. The method does also have some special applications which aim to solve certain problems less easily tackled by the reflection method; furthermore, combined refraction/reflection surveys can often be achieved at marginally extra cost if surveyed concurrently with reflection surveys, and adoption of this exploration strategy may find wider acceptance in the future with consequent development of better acquisition techniques and interpretational expertise.

Some of the purposes for which refraction surveys are conducted at present are:

1. Basement mapping for reconnaissance purposes.

2. Horizon mapping beyond the depth range of the reflection method.

3. Provision of control data for reflection interpretation to give both depth and velocity structure; such data are of particular value in areas where borehole control is sparse or absent.

4. To map horizons which underlie complex geological structures such as salt-domes, shale and salt diapirs, igneous intrusions and the multiple thrust-faults of orogenic belts; often a combination of refraction survey and reflection undershooting is used to tackle such problems.

5. To determine the thickness of low velocity unconsolidated or weathered layers encountered in many land exploration terrains. Data so obtained are used to apply corrections during the processing of reflection seismics (see chapter 3).

Refraction of seismic waves is governed by the simple law, Snell's law, illustrated in figure 7/5 which states $V_1/V_2 = \sin\alpha/\sin\beta$ where V_1 and V_2 are the seismic velocities either side of a boundary plane. Refraction exploration depends on an effect whereby acoustic energy is propagated along the plane of the boundary, with emission of energy into the media both sides of the boundary. Seismic waves so propagated are termed head waves. The effect occurs under the condition that $\beta = 90°$, in which case α is said to be the critical angle α_c such that $\alpha_c = \sin^{-1} V_1/V_2$. In the simplest type of refraction shooting, reversed in-line profiling, an interpreted section can be built up which should correlate well with a seismic reflection profile, but

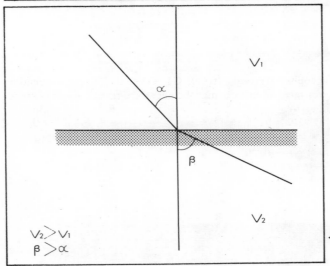

with the advantage that refracting horizons are defined in terms of depth, not two-way reflection time. Other types of refraction shooting include fan-shooting and broadside shooting (see later).

In practice, the methods of analysing and processing refraction data can be quite complicated unless the geological conditions are fairly simple, ie, horizontal layers of isotropic rock media and in the absence of velocity inversions. Often, sites can be selected in a prospect area, or field programmes designed which can minimise actual deviations from these conditions, particularly in reconnaissance work or in situations where the principal aim is to gain data on velocity information and depth control to aid reflection interpretation. For the interpretation of refraction data over complex structures computer methods have been

Figure 7/5: Refraction of seismic waves according to Snell's Law

Figure 7/6: Refraction of seismic waves in a three-layer earth.

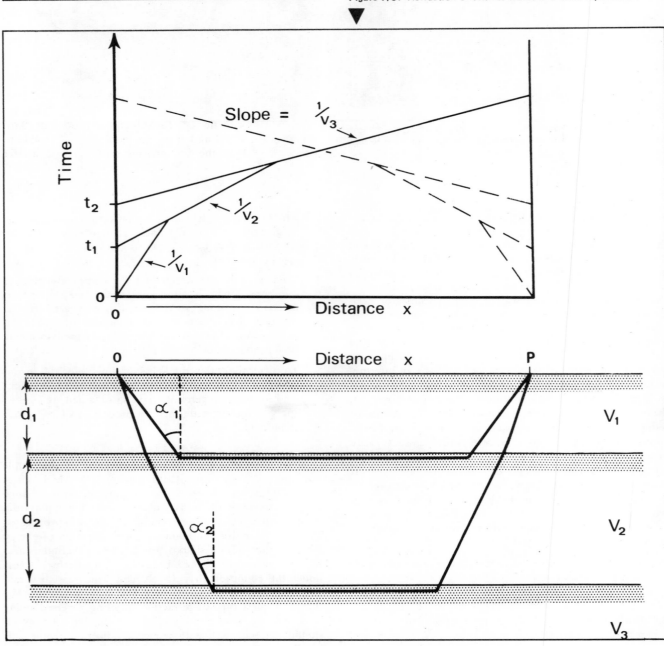

developed, as well as graphical aids.[†] Here, only a very simple case will be treated, that of a simple three layer structure as shown in figure 7/6 where $V_1 > V_2 > V_3$ and the uppermost two layers have a thickness d_1 and d_2. Let us assume a shot-point at position 0 with receivers (groups of geophones or hydrophones) spaced at intervals along the x axis. In refraction exploration, it is the time between shot instant and first arrival of acoustic energy through the rock layers which is of principal use. Receivers close to the shot-point will receive, as first arrivals, waves which has passed directly through the uppermost layer, velocity V_1, these first arrivals being followed by reflections from deeper layers. Receivers, at increasing distance from the shot-point will receive, as first arrivals, seismic waves refracted as head waves of velocity V_2 or of V_3, depending on distance along the x axis. At the more distant receivers, arrivals from layers 2 and 1 will follow the first arrival refracted waves from layer 3, along with reflected waves, multiples and other refractions, giving a complicated signal. However, with processing, it is usually possible to detect and display the primary signals from the upper layers, as well as the first arrival signal. A time distance graph of first arrival times has the characteristics shown in figure 7/6. From this graph it is possible to extract data on both the seismic velocities and thicknesses of the rock layers. Slopes of straight line segments of the graph can be measured, these being equal to $1/V_1$, $1/V_2$, and $1/V_3$.

If the intercepts of these lines through the time axis are plotted, then the first segment passes through $t = 0$, the second segment through t_1 such that:

$$d_1 = \frac{V_1 t_1}{2 \cos a_1}$$

and the third segment through t_2 such that:

$$d_2 = \frac{V_2 \left[t_2 - t_1 \dfrac{\cos a_3}{\cos a_1} \right]}{2 \cos a_2}$$

where,

$a_1 = \sin^{-1}(V_1/V_2)$, $a_2 = \sin^{-1}(V_2/V_3)$ and $a_3 = \sin^{-1}(V_1/V_3)$

These relationships can be adjusted and extended for dipping interfaces and for more than three layers, but with increasing complexity of the mathematical relationships. A range of interpretation methods are discussed in *Seismic Refraction Prospecting*,[†] including those using graphs and nomographs, as well as applications of the delay-time method and wavefront methods. These methods include the treatment of dipping layers and internal velocity gradients.

A profile as described above, will not by itself detect the presence of dipping events or other deviations from the simple horizontal layer case. If however the profile is reversed, that is, if the shot-point is removed to a point P, and the same spread of receivers is used, then an exactly symmetrical time-distance graph should result; the dashed lines in figure 7/6. Any deviation from this symmetry can

be used as a measure of dip between layers.[‡]

In practice, surveys are usually designed so that data acquisition provides a set of overlapping reversed profiles. A typical layout for a long land profile is shown in figure 7/7. Shot-holes are positioned at SH0, SH1, SH2 etc, and geophone groups at G0, G1, G2. . . G50 etc. Depending on target depth, it may be necessary to use spreads extending to 40—50km or more from each shot-point. Considering the shot, at SH2, it can be seen that both up-dip (to the east) and down-dip data (to the west) will be observed from this shot position.

In figure 7/8 the system is shown which is commonly adopted for refraction surveys at sea. The seismic source can be either a conventional reflection seismic source, such as an airgun or gas-gun array, or, in the case of long-range experiments it is often necessary to use explosive charges at distances beyond that at which good arrivals can be detected using the relatively low energy reflection sources. In the system illustrated a sonobuoy is used at a receiver-transmitter; the hydrophone can be either suspended in the sea from a free-floating or anchored buoy, usually on an elastic suspension designed to dampen out sea wave motion, or it can be laid on the seabed attached to an anchored buoy. The buoy contains an amplifier and radio transmitter which transmits seismic signals to a shipboard receiver. Acoustic signals are displayed either on a multi-channel oscillograph or on a graphic recorder as well as being recorded on magnetic tape for subsequent processing. Shot instant is recorded, and from the display it is possible to detect, as well as the first arriving seismic signal, a later strong high frequency signal which is the acoustic impulse which has travelled directly through the sea water from source to hydrophone. Thus measurements can be made of interval times: shot instant to first arrival and shot instant to direct-wave arrival. This second time interval divided by the velocity of sound through sea water gives the distance between shot-point and sonobuoy. Such a measurement is particularly important if the sonobuoy is free-floating and is likely to be drifting in the sea current or by force of wind in which case its position will vary over the duration of the profile acquisition period. An alternative marine acquisition system which is much more costly in use but which can give much higher quality data utilises a two-ship system. In this case a conventional digital reflection acquisition system can be used, appropriately adjusted to accept low frequencies. A survey with such a system might be conducted as follows; a 2.4km (48-channel, 50m section interval) cable would be towed away from a fixed shooting location, a shot being fired at the beginning of every 2.4km of travel along the profile line by the recording vessel.

---

† A.W. Musgrave, *Seismic Refraction Prospecting* (Society of Exploration Geophysicists, Tulsa, Oklahoma, 1967).

‡ For example, see M.M. Slotnik, 'Agraphical method for the interpretation of refraction profile data' *Geophysics*, 15 (1950), pp. 163—180.

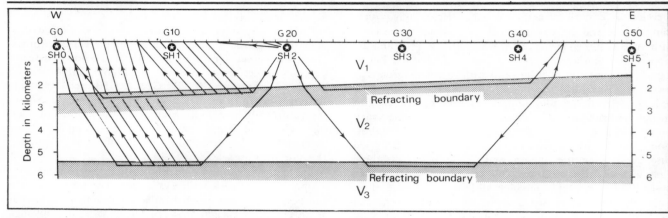

Figure 7/7: Layout of reversed in-line refraction survey on land.

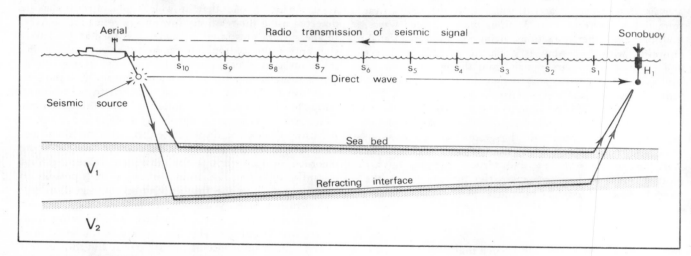

Figure 7/8: Layout of seismic refraction experiment at sea.

In practice a number of shot locations might be used, with the recording ship sailing a profile up to, then away from, each shooting location. Thus a series of overlapping reversed profiles is built up.

In figure 7/9 an example is shown of an interpretation of the results of a marine refraction experiment carried out in the Irish Sea. Results indicate a layer of approximately 1.5km thickness of velocity 3.46km/s overlying rocks of velocity 5.07km/s. Geologically, these results suggest that 2km of Permo-triassic rock overlie a Palaeozoic basement.

In figure 7/10 a refraction seismic section is shown from a land seismic survey. On this section the high velocity basement refracter (γ) crosses a fault. Also shown is the interpretation of a set of travel time curves using the wavefront method over 45km of profile. In this particular study, velocity information for rocks overlying the refractor (γ) was based on reflection seismics, though on reception, low velocities can be detected as both first arrivals and as late refracting arrivals in the 4—7s interval. The application of refraction seismics to an investigation of a complex subsalt tectonic problem in northern Germany is illustrated in figure 7/11. These travel time curves and the depth presentation are from part of a 300km refraction line, the presentation giving a good example of the use of Thornburgh's wavefront method.* Other refraction seismic techniques include fan-shooting, broadside (arc) shooting, radial shooting and combined reflection and refraction

surveying. Fan-shooting is a well established technique which was devised principally to detect and map salt domes within a relatively low velocity section. Firstly, a calibration profile is established in the area of study where the seismic section is known to be normal, ie no salt domes are present. Then a number of fan-arrays are shot (see figure 7/12) and early arrivals (time-leads) are plotted which indicate where ray paths have passed through the high velocity salt domes.

Broadside (arc) shooting is similar to fan-shooting but has more general application. A central shooting location is used and detector stations are located on circle segments. Different radii may be used, values being gauged so that the various refractors being investigated will provide first-arrival signals in the sections produced. An example of broadside shooting is shown in figure 7/13 where the object of study was fault detection at depths where reflection data might provide inconclusive evidence. Radial shooting is similar to arc shooting but in this case a geophone is lowered into a well to the level of an identified refractor, then a series of shots are fired in an arc to investigate structure associated with the refractor.

*H.R. Thornburgh, 'Wavefront diagrams in seismic interpretation' *Bull AAPG* 14 (1930)' pp. 185-200

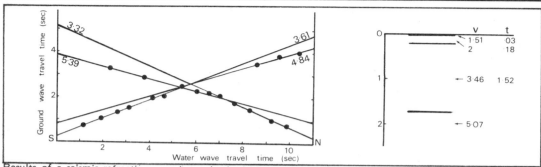

Figure 7/9: Results of a seismic refraction experiment in the Irish Sea (from M. Bacon and R. McQuillin. 'Refraction seismic surveys in the North Irish Sea'. *Jl. Geol. Soc. Lond.*, 128 (1972), pp.613—21).

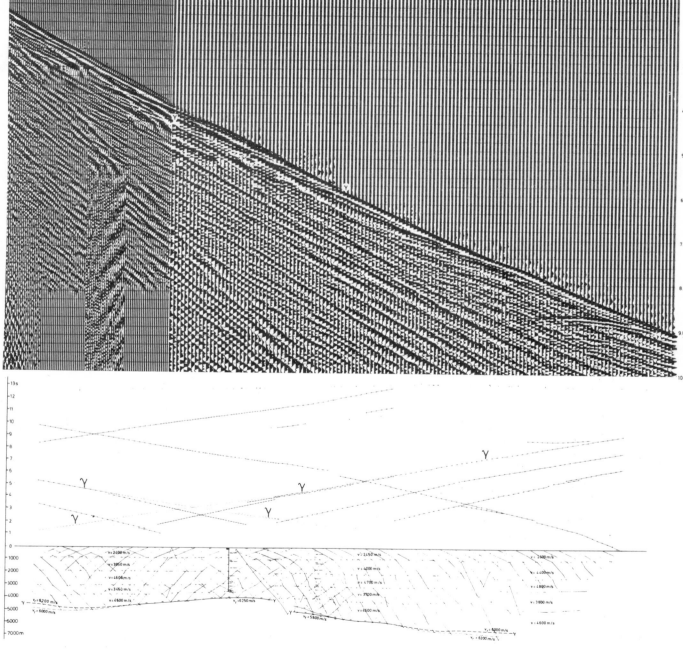

Figure 7/10: Refraction seismic section (above) with fault indication in the basement refracter (γ). Travel time curves (below) of a 45km long refraction line with depth interpretation of the basement refractor (γ) using the wavefront method.. (*Courtesy: Prakla—Seismos*).

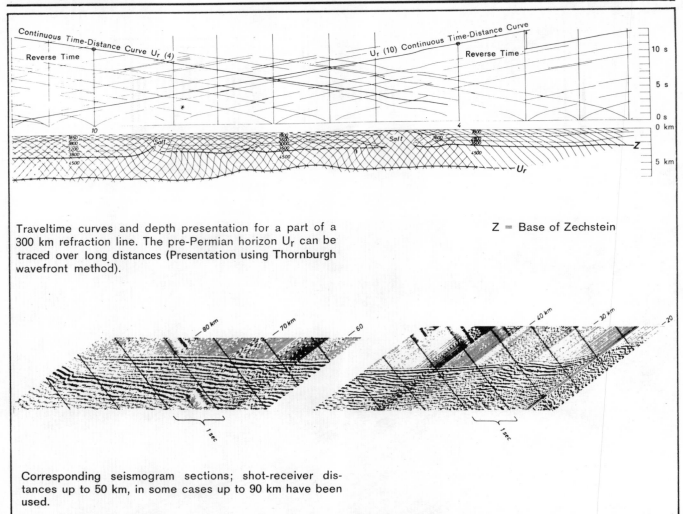

Traveltime curves and depth presentation for a part of a 300 km refraction line. The pre-Permian horizon U_r can be traced over long distances (Presentation using Thornburgh wavefront method).

Z = Base of Zechstein

Corresponding seismogram sections; shot-receiver distances up to 50 km, in some cases up to 90 km have been used.

Figure 7/11: Application of refraction seismics to a subsalt tectonic problem in a deep saltdome basin in Northern Germany. (*Courtesy: Prakla—Seismos*).

Figure 7/12: Layout of refraction fan-shooting survey Profile A—C is shot as a reversed refraction profile to establish a calibration of the normal time-distance curve for the area. Shaded area shows approximate outline of saltdome responsible at depth for the plotted time leads (after L.L. Nettleton, *Geophysical Prospecting for Oil,* p.227 McGraw—Hill, New York, 1940, p.277).

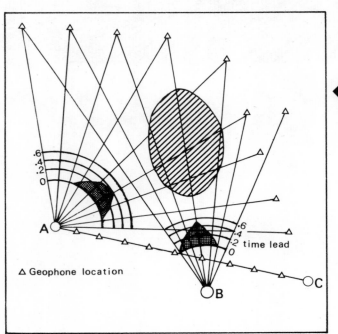

△ Geophone location

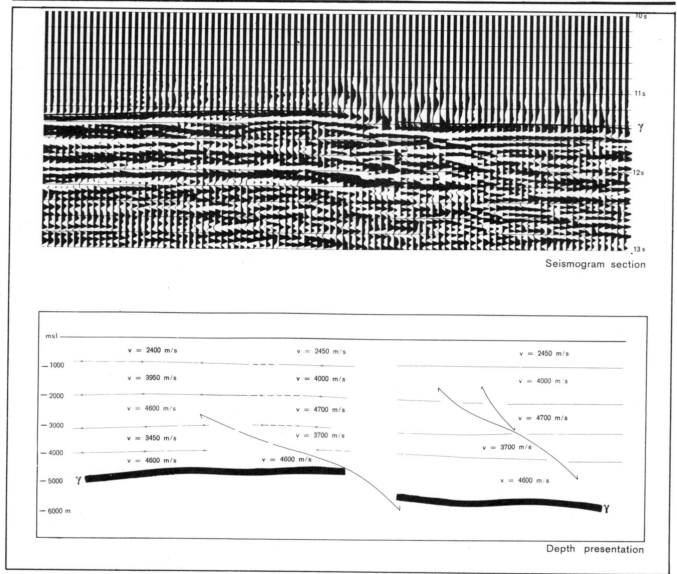

Seismogram section

Depth presentation

Figure 7/13: Results of a broadside-refraction survey. Offset between shots and receiver = 5km.
(*Courtesy: Prakla—Seismos*).

An interesting application of combined reflection/ refraction surveying has been described by Prakla-Seismos.* The method depends on utilising geophone spreads already in place during reflection profiling and by additional shooting obtaining broadside sections. An example of the method from a survey carried out in the South German Molasse Basin is shown in figure 7/14. The method can only be applied if a good refractor is present. In the example shown, the location of a fault is traced from its position on the left-side refraction offset line, through its position on the reflection profile, to its position as indicated on the right-side refraction offset line. The method can under some circumstances give a more positive indication of faulting than is seen on the reflection record as well as giving a determination of fault strike. An example of a section showing fault detection by this method is shown in figure 7/15.

7.3 Magnetic methods
An aeromagnetic survey is an economical method of undertaking a geophysical reconnaissance of any relatively unexplored region by providing data on broad scale structural trends, the positions of faults, the distribution of shallow and deep crystalline basement, as well as the occurrence and distribution of volcanic rocks within sedimentary basins. Such information can be invaluable in planning more costly seismic reflection exploration work which will aim to detect defined economic targets. Magnetic survey results are furthermore useful to the seismic interpreter even at the stage of detailed interpretation of a close grid of seismic lines. He may have access to both the results of an aeromagnetic reconnaissance survey

*G. Badtke 'Better co-ordination of faults by combined reflection and refraction surveys' *Prakla Seismos Report* No. 1/76 pp. 10-11.

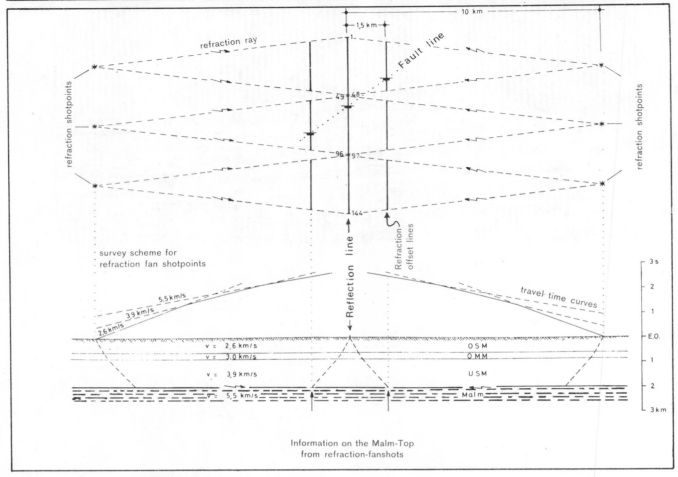

Figure 7/14: Combined reflection-refraction survey. Any fault which displaces the Malm refracting horizon will be detected in both the reflection section and in the refraction offset lines. (*Courtesy: Prakla—Seismos*).

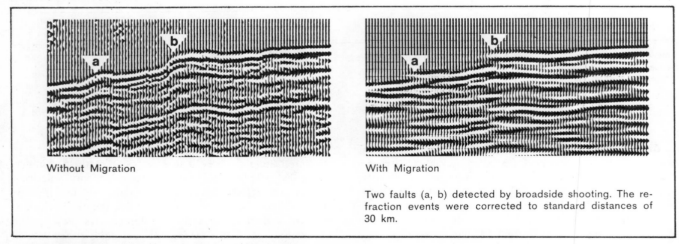

Without Migration

With Migration

Two faults (a, b) detected by broadside shooting. The refraction events were corrected to standard distances of 30 km.

Figure 7/15: Detection of faults by broadside-refraction shooting. This example shows how migration processing can be applied to refraction data to improve definition of the fault structures. (*Courtesy: Prakla—Seismos*).

as well as magnetic profiles surveyed along actual seismic lines, whether on land or at sea. We are mainly concerned here with how such data can be used by the interpreter. Later, in chapter 9 the Moray Firth case history illustrates how the results of an aeromagnetic survey can be used as an aid to seismic interpretation, particularly in the early stages of exploration of an area when seismic coverage is open and little or no drilling results are available.

The basic principle underlying magnetic surveying is that some rock types, notably, though not exclusively, of igneous origin, are naturally much more highly magnetised than

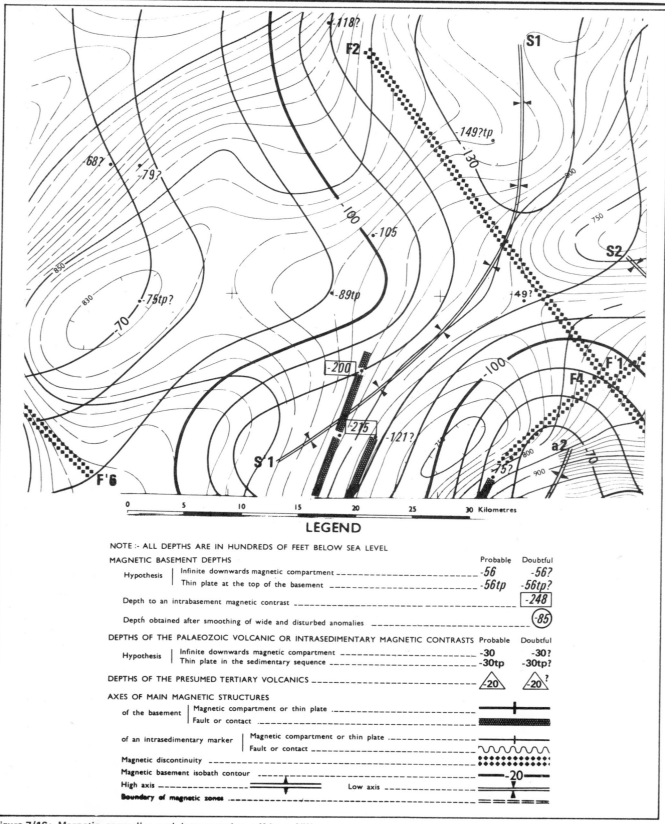

Figure 7/16: Magnetic anomalies and interpretation offshore NW Scotland.
(*Part of an aeromagnetic survey by Hunting Geology and Geophysics, UK*).

others. These bodies of magnetised rock produce a magnetic field which is apparent as a local perturbation of the earth's magnetic field.

Magnetic surveys therefore aim to detect anomalous magnetic field variations associated with geological structure; magnetic anomalies are determined by comparing measured magnetic field values with either a local or whole-earth reference field. Most modern survey work uses equipment which measures the total field; the earth's main field varies from approximately 0.4 oersted at the equator to 0.6 oersted at the poles. The unit of measurement in most common use in magnetic exploration work is termed the gamma and 1 gamma is equal to 10^{-5} oersted. Most present day surveys are referenced against the International Geomagnetic Reference Field (IGRF), which is defined in terms both of the main field and of secular variations. Thus, for any locality on the earth's surface, and for a particular epoch, a reference field value can be calculated. Secular variations are sufficiently small for it to be acceptable to assume no significant secular change over the period of a normal survey programme (say up to 6 months). A map showing contoured values of the differences between magnetic field measurements and the reference field values is called a magnetic anomaly map. Magnetic profiles are sometimes shown on top of seismic sections and anomaly values for these are usually similarly derived. Figure 9/3 shows a magnetic anomaly map of the Moray Firth area. This map is derived from the results of an aeromagnetic survey flown at 1000ft above sea or ground level with a line grid spacing of 2km x 10km. Instrument accuracy is approximately 1 gamma and the results are usually contoured at a 10 gamma interval.

Geological interpretation of magnetic data usually aims as a first objective to indicate the depth of burial of magnetised rocks. Major geological features such as faults, anticlinal and synclinal axes may also be determined. As a further refinement, model structures are derived with computed magnetic fields which give a good fit to the observed magnetic anomaly fields. These geological models can then be compared with results of other geophysical surveys and assessed in terms of geological probability.

Interpretation maps are often produced by aeromagnetic survey companies which are based on automated or semi-automated procedures for handling the large quantity of data involved. Fourier analysis techniques are widely used; short wave-length anomalies are produced by magnetic rock bodies occurring near ground level whereas long wave-length anomalies originate at deeply buried levels. In figure 7/16 part of such a map is shown. It can be useful in seismic interpretation to compare such a map with interpreted isochron or isopach maps for the same area. If the seismic data allow mapping of depth to economic basement, it would be interesting to see if this correlates well with depth to magnetic basement. The magnetic map will also indicate if volcanic or intrusive igneous rocks occur within the mapped area. Layers of high velocity volcanic rocks can in some circumstances appear similar to evaporites in a seismic section. Volcanic rocks are usually moderately magnetic, evaporites are not, thus it is possible to differentiate between the two possibilities. Similarly, it can be important to differentiate on seismic sections between ridges or block-faulted horst structures of normal basement rock and igneous intrusive structures such as volcanic necks and vents, dyke swarms etc; a differentia-

tion which can be greatly aided by a study of and interpretation of magnetic data.

Two examples of the use of magnetic data as an aid to the interpretation of seismic results are discussed now and in both cases a qualitative interpretation of the magnetics is sufficient. The first example is derived from the Sea of the Hebrides, NW British Continental Shelf. In this area Mesozoic sediments are in places overlain by high velocity (presumed Eocene) lavas. Because of generally poor reflector quality and the presence of numerous igneous intrusives, seismic data quality is poor. Economic basement in the area comprises Pre-Cambrian and Lower Palaeozoic rocks. Prior to the survey, Tertiary sediments of post-Eocene age were unknown in this area; however, seismic sections indicated in some localities up to in excess of 1s two-way time of low velocity sediments with a character unlike that of the Mesozoic section. A post-lava age was confirmed by showing that in at least one locality, thick sediments of this type overlay a thickness of Eocene lavas. In figure 7/17 two seismic sections are shown over one of which a marine magnetometer profile was obtained. Also shown is an interpreted magnetometer profile across the axis of the basin which has also been mapped using shallow profiling (sparker) equipment. Figure 7/18 shows eventual interpretation of Line 1.* In this case it was the interpretation of a combination of conventional seismic data, seismic profiling data, magnetic data and seabed sampling which led to a plausible and scientifically significant geological interpretation. The particular contribution of the magnetics was to indicate and allow mapping of lavas at seabed, such areas being characterised by short wavelength, high amplitude anomalies. Also, the anomaly patterns of Lines 2 and 5 are more consistent with the interpretation of a basin underlain with lavas than with a possible alternative interpretation of the seismic data, that the area with lavas absent at seabed represented an area where these had been removed by subsequent erosion in the Quaternary and deposition of superficial sediments on top of the Mesozoic.

As a second example of integrated interpretation of seismic and magnetic data, a structure is illustrated which was discovered during an early (1968) seismic survey of the southern Irish Sea. The seismic section (figure 7/19) indicated a strongly positive feature which was difficult to interpret as it was only crossed by one seismic line in a very open survey grid. No other seismic or drilling data were available in the area and a number of interpretations were possible: that the structure was associated with an igneous intrusion; that it was associated with a basement ridge; or, that it was a diapiric salt structure. The fact that the structure had no associated magnetic anomaly strongly suggested that the latter interpretation was correct.

The mathematical basis of quantitative interpretation and model fitting techniques are beyond the scope of this book; standard geophysical texts are listed as suggested reading which should be consulted if the reader wishes to undertake such quantitative interpretation. As a very approximate guide to estimating depth of burial of magnetic rock bodies, the interpreter can however abstract profiles normal to the trend of mapped magnetic anomalies, then measure

---

*This example is based on the work of D.K. Smythe, and N. Kenolty, 'Tertiary sediments in the Sea of the Hebrides' *Jl. Geol. Soc. Lond.* 131 (1975) pp. 227–233.

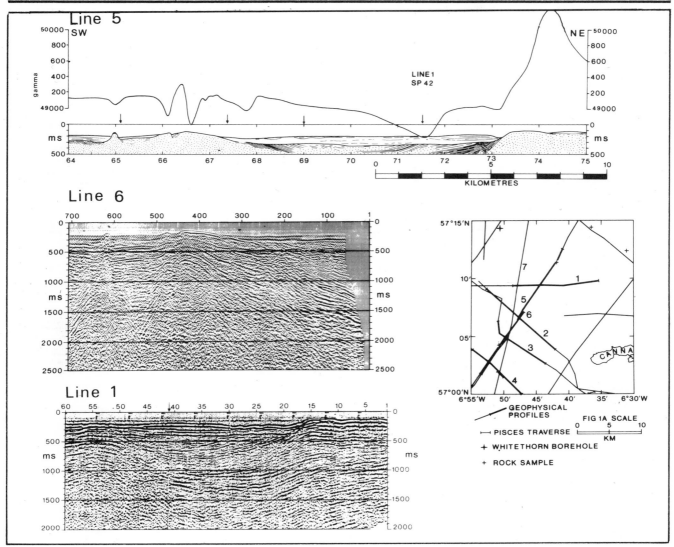

Figure 7/17: Geophysical study of a small basin of presumed Tertiary sediments in the Sea of the Hebrides. Map shows positions of geophysical profiles, a Pisces submersible traverse, shallow borehole locations and rock sample sites. Line 5 was surveyed using magnetometer and sparker; the sparker interpretation is shown beneath the magnetic profile. Line 6, a 12-fold stacked seismic section overlaps the south-western part of line 5 as shown. Line 1, a 24-fold stacked section (shot by Forest Petroleum UK) intersects the basin, and an interpretation of this section is shown in figure 7/18.

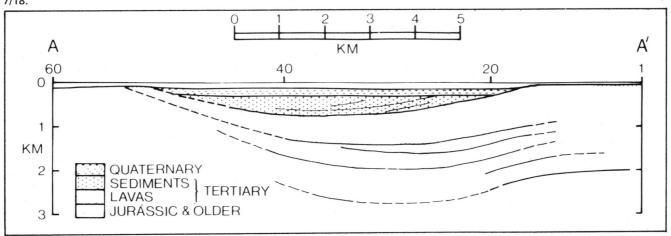

Figure 7/18: Interpretation of seismic line 1 of figure 7/17.

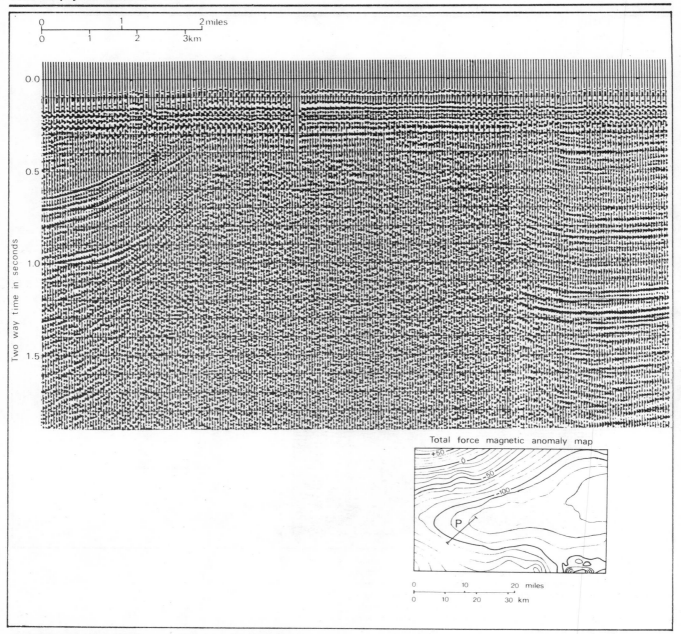

Figure 7/19: Seismic section from the Cardigan Bay area of the southern Irish Sea. Magnetic map (inset) shows absence of a magnetic anomaly associated with the high structural feature, marked P. Magnetic contours are at 10 gamma interval.

the half-amplitude width of prominent deflections. This horizontal distance equates to a maximum depth value to the source of the magnetic anomaly. To illustrate the effect, computed magnetic anomalies over two simple structures are shown in figure 7/20 with varying depths of burial; the size of the magnetic body and its magnetisation are kept constant. For the vertical cylinder example, the half-amplitude width varies from 3km at 2km depth burial to 7.5km for 5km depth burial; in all cases an over estimate.

Application of this method to an actual interpretation is illustrated in figure 7/21 and 7/22. Figure 7/21 shows an aeromagnetic anomaly map of the North Minch area of northwest Scotland. This map shows an area of low magnetic gradients intersected by a strong linear magnetic feature. Geological interpretations suggest that a Mesozoic basin

lies between the landward basement areas of the Outer Hebrides and the mainland of Scotland. The linear magnetic feature is assumed to be associated with an igneous dyke of presumed Tertiary age, one of a large suite of dykes which were intruded in the early Tertiary during a period of intense igneous activity in northwest Scotland. Figure 7/22 shows two marine magnetic profiles across the feature compared with seismic reflection profiles. Study of the magnetic profiles indicates a depth to top of the dyke of 1600m in profile A and 2700m in profile B. These figures compared with depth values of 1750m and 3150m obtained from the seismic sections using a velocity of 2000m/sec for the sediment cover under which the dyke is buried.

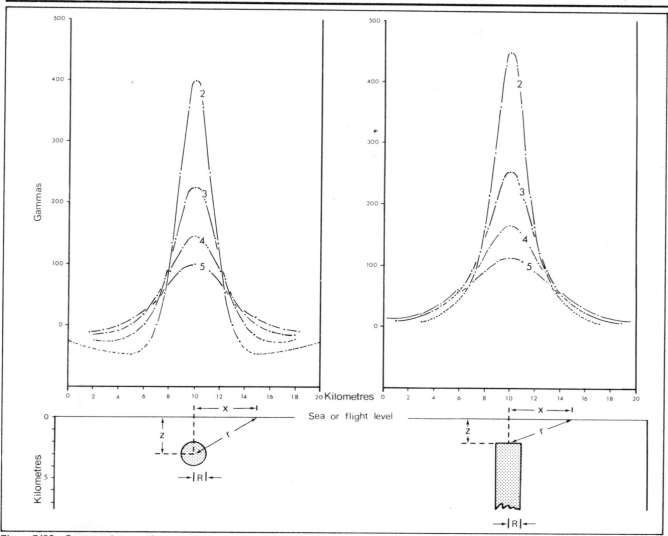

Figure 7/20: Computed magnetic anomalies over simple structures. On the left, over a horizontal cylinder, on the right over a vertical cylinder. Curves 2–5 represent different depths of burial (Z) from 2–5km.

7.4 Gravity methods

Gravity surveying can, in many respects, play a similar role to that of magnetic surveying within any exploration programme based principally on the interpretation of the reflection seismic data. Gravity surveys are often made as a means of obtaining, at relatively low cost, a geophysical reconnaissance of any relatively unexplored region. On land, the method is fairly straight-forward and operations utilise compact easily transported instruments which are simple to use. At sea, the situation is more complex, and the whole operation of gravity surveying demands very much higher technical resources, both in terms of man-power and instrumental hardware. Attempts have also been made to survey gravity from aircraft (particularly heli-copters), but it is only very recently that systems have been developed which are able to monitor craft position, course, velocity and vertical acceleration to an accuracy which allows gravity computation at a level of accuracy which is acceptable for reconnaissance survey work. The results of regional gravity surveys, both on land and off-shore, allow delineation of boundaries of sedimentary basins as well as gross estimates of total thickness of infill within such basins. Combined interpretation of gravity and magnetic data can give improved reliability in such estimating. Thus areas can be selected on the basis of regional, gravity and magnetic surveys for further investigation by the reflection seismic method. However, our main concern here is to consider how the interpreter of seismic data can use gravity results at that stage of his interpretation where he is using a grid of seismic lines in a, by then, defined prospect.

The laws of gravitational attraction are similar to those of magnetic attraction; in both cases we are concerned with potential field theory. In gravity studies it is Newton's law of gravitation which describes the nature of attraction between bodies of matter. This force of attraction is expressed in the equation:

$$F = \frac{G m_1 m_2}{d^2}$$

where F is the gravitational force, d is the distance between the point masses, m_1 and m_2, and G is the gravitational

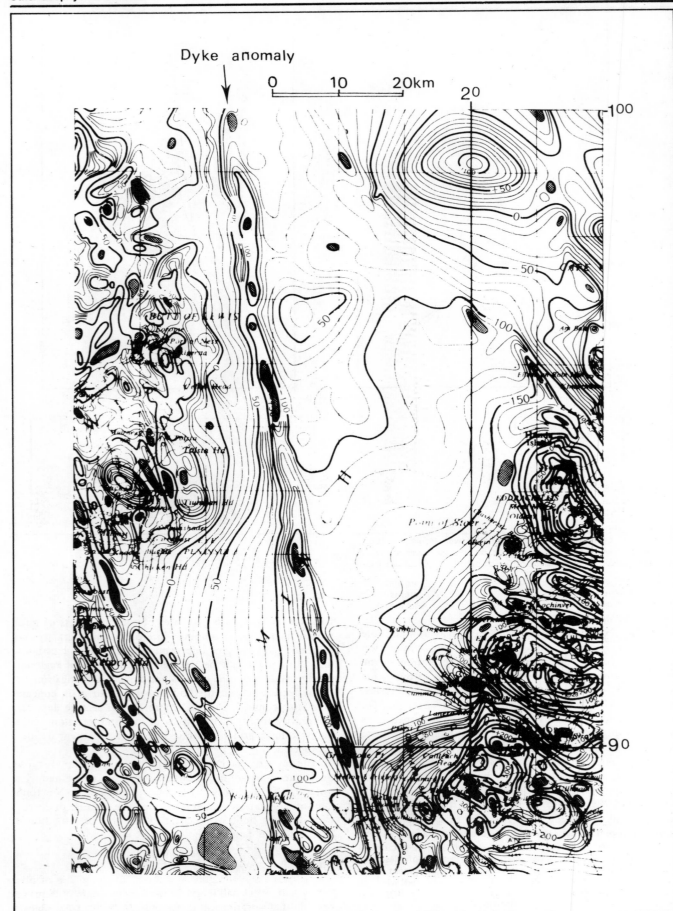

Dyke anomaly

0   10   20km

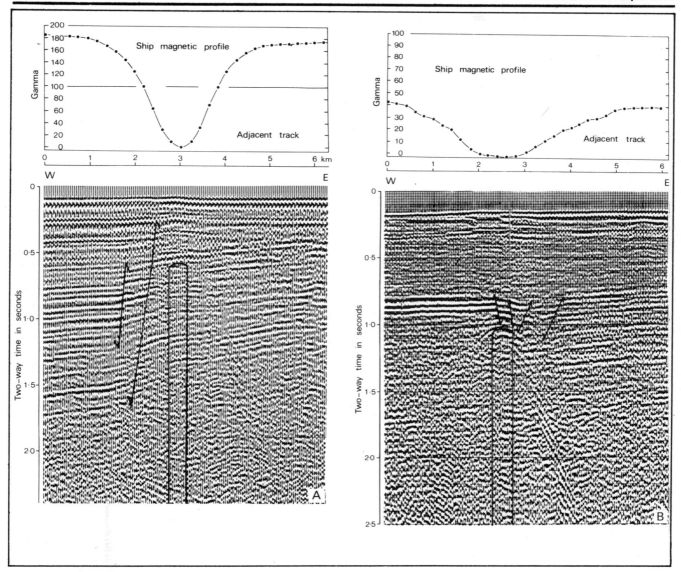

Figure 7/22: Comparison of magnetic with seismic profiles from the North Minch. Note single-point type diffraction patterns due to reflections from top and margins of dyke intrusion, also the minor faulting associated with dyke emplacement.
(Seismic profile A, courtesy IGS; seismic profile B, courtesy BP and Gas Council; marine magnetic survey by IGS).

Figure 7/21: Total force aeromagnetic map of the North Minch area off NW Scotland (between the Scottish mainland and the Outer Hebrides). High frequency large amplitude anomalies over land areas are associated with Pre-Cambrian basement. The sea area is characterised by generally low amplitude anomalies and gentle magnetic gradients due to the offshore development of Mesozoic basins underlain by Torridonian sediments. The prominent near-linear anomaly which intersects the basin is associated with an igneous dyke, probably of Tertiary age. (Adapted from IGS aeromagnetic map).

constant which has been experimentally determined as $6.670 \times 10^{-8} \ cm^3/g \, s^2$.

As with magnetic fields, we are dealing with an inverse square law of attraction. If the earth were a perfectly homogeneous non-rotating sphere in empty space, then the gravitational pull would be uniform over its entire surface. This is not the case; the earth is not a sphere but an ellipsoid of revolution; a site at either pole is closer to the earth's centre than one at the equator. Also the earth's revolution leads to an acceleration force of an identical nature to the force of gravity, which is also an accelerational force. The sum of these main effects on variation of gravity over the earth's surface, as measured at sea-level, is that gravity at the earth's poles is approximately 0.5 per cent higher than at the equator. A reference field can be defined for the earth which takes account of these effects and that used in modern gravity survey work is the 1967 International Gravity Formula.

From the results of gravity surveys anomaly values are computed (as with magnetic surveys) which are in this case the differences between calculated reference gravity field values and observed gravity values. The anomalies of interest to us have small length scales (say less than 100km) and are due to variations in density of the rocks from place to place within the crust. Anomaly maps and profiles can be used to study geological structure involving rock units which exhibit contrasting rock density. In land surveys other corrections are applied for topographical effects; it is necessary to correct for height changes between observation sites and the values are usually computed to a sea-level datum.

Allowance is made for variation in height of observation (the Free-Air correction), the gravitational effect of the layer of rocks between the observation site and sea-level (Bouguer correction) as well as for the gravitational effects of surrounding hills and valleys (the terrain correction). For further details see one of the standard geophysical texts listed as suggested reading. After all such corrections are applied, anomaly values are contoured and/or profiles prepared. Such maps or profiles (in terms of Bouguer anomaly values) are in exploration work usually expressed in milligals. One milligal is 1/1000gal which is an acceleration of $1cm/s^2$. The earth's main (reference) field varies from approximately 978gal at the equator to 983.2gal at the poles. Land surveys are made to an accuracy of better than 0.1mGal. and results are contoured at one milligal intervals. Measurements at sea (except where seabed gravity meters are used, and this is a special application of marine gravity work in which data is treated in a way almost identical to land data) are made using instruments mounted on stabilised platforms linked to computer control systems which aim to compensate and correct the accelerations brought about by the ship's motion.[*] Accuracies obtainable are less than for land surveys. A good marine survey may have an average cross-tie error of approximately 1 milligal and results are usually contoured at either 5 milligals or, if the survey has been exceptionally well controlled, at 2 milligal intervals. Surveys made from aircraft or helicopters are likely to suffer even greater inaccuracies especially in application of corrections, and an anomaly accuracy of 2—4 milligal is probably the best achievable with present-day technology.

Before any attempt can be made to interpret the patterns of gravity anomalies associated with a particular area, some estimate must be made of local contrasts in rock density between different structural units. Using such estimates of density-contrast, model geological structures can be tested to find the most plausible disposition of rock units which will produce , by computation, the same anomalous gravity field pattern as has been observed in the actual gravity survey. This process of interpretation is well illustrated in the Moray Firth case history in chapter 9 of this book. In table 7/1 a list is shown of the densities of some common rocks. It can be seen that some common rock types exhibit a fairly large range of densities, one of the more difficult aspects of gravity interpretation is due to the fact that even low-porosity crystallised basement rocks exhibit a similar wide density range due in this case to differences in mineralogical composition. Average density of the earth's upper crust is approximately 2.7gm/cc. Thus it can be seen that large granite intrusions (density 2.55—2.65gm/cc) will cause reduced gravitational attraction; that is, a negative gravity anomaly or gravity low. Similarly a massive basic intrusion or a thick layer of basalt lavas will cause a gravity high.

[*]For a further discussion see R. McQuillin and D.A. Ardus, *Exploring the geology of shelf seas* (Graham & Trotman, London, 1977).

Table 7/1: Common Densities

	g/cc
Oil	0.90
Fresh water	1.00
Sea water	1.03
Unconsolidated sand	1.95—2.05
Boulder clay	1.90—2.10
Porous sandstone	2.00—2.60
Rock salt	2.10—2.40
Granite	2.55—2.65
Quartzitic sandstone	2.60—2.70
Compact limestone	2.60—2.70
Gneisses	2.70—3.00
Basalt	2.70—3.10
Basic intrusive rocks	2.80—3.20
Ultrabasic intrusive rocks	2.80—3.30

As an aid to seismic interpretation, gravity data are mainly used in direct comparison with seismic sections to test the plausibility of a particular seismic interpretation wherever this is problematical. Fault structures can be modelled to test seismic correlation across major discontinuities. Estimates of the density of seismic basement can sometimes be used to interpret its nature; for example, a seismic basement horst might consist of Devonian sandstones of relatively low density (a potential oil or gas reservoir) or a non-porous basement of higher density (a structure with no reservoir potential). Gravity modelling can be used to test such alternative hypotheses. Rock salt has a relatively low density, and large structures involving thick salt intervals will have associated gravity anomalies. Salt is also a high seismic velocity material and in areas of complex salt movements, it is often the case that reflection data show only poor penetration beneath salt layers. Again, gravity modelling can be used to test a range of possible seismic interpretations where these are based on unreliable seismic evidence. Another use of gravity data is in the study of deep structure within sedimentary basins to depths beyond that penetrated by available seismic data. In such situations it is possible to evaluate the three-dimensional gravitational effect of the entire sedimentary infill as mapped from a seismic interpretation. This gravitational effect can then be compared with the observed gravity field to obtain an indication of the main elements of underlying structure. In this way, structural features can be detected which merit further study by the seismic method now optimised for deep penetration in these areas of maximum interest.

To illustrate the analytical method of gravity interpretation we shall use here a fairly simple example. The formula from which a gravity profile across a fault or step structure can be calculated is given by:

$$\Delta g = 2G \cdot \Delta \rho \left[x \cdot \ln r_1/r_2 + D(\pi - \theta_2) - d(\pi - \theta_1) \right] \text{ mGal},$$

where $\Delta \rho$ is the density contrast and d and D are the depths to top and bottom of the step, ($D-d$ being the throw of the fault if it is a simple fault structure; see figure 7/23 for definition of r_1, r_2, θ_1, θ_2, and x). It can be seen from the above that the total gravity change across the structure is given by:

$$\Delta g = 2\pi G \cdot \Delta \rho \ (D-d) \text{ mGal}.$$

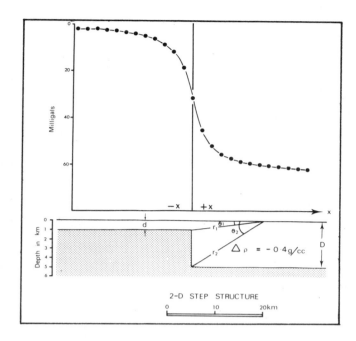

2-D STEP STRUCTURE

Figure 7/23: Calculated gravity profile across a model step structure.

Figure 7/24: Comparison of gravity profile and seismic section across a major fault (the Great Glen Fault) in the Moray Firth, off NE Scotland.

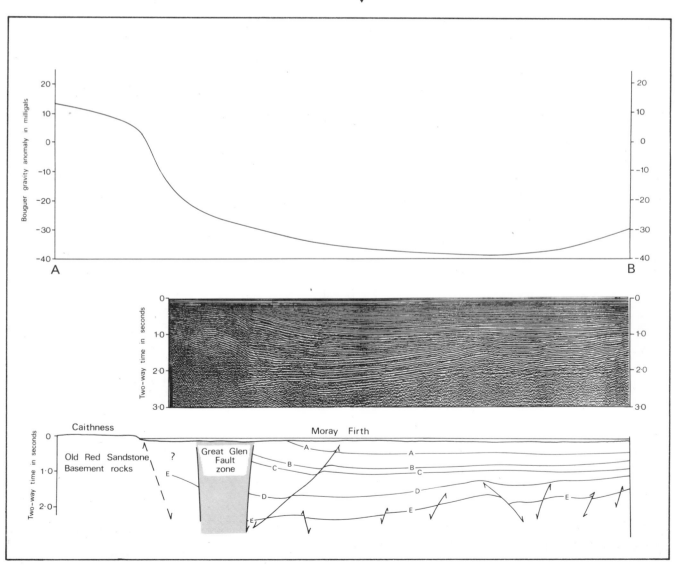

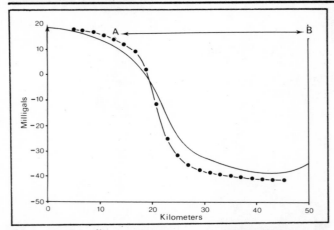

Figure 7/25: Computed (dots) and measured gravity profiles across margin of the Moray Firth sedimentary basin. Model is a 4km step identical to that illustrated in figure 7/23.

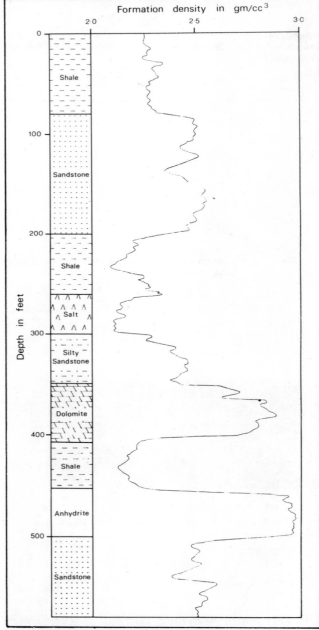

Figure 7/26: Schematic density log.

That is, for the structure illustrated in figure 7/23, a gravity change of 76 milligals. In figure 7/24 the direct comparison is shown between a gravity profile across a major fault in the Moray Firth and a seismic section across the profile, and in figure 7/25 a gravity profile is fitted to a model fault structure. The relationship between seismic interpretation and gravity interpretation is clearly demonstrated.

As was stated early in this discussion, it is important to have knowledge of density contrasts associated with geological structures being interpreted. One of the most common methods of determining rock density values is by laboratory measurement of the saturated density of rock samples, either collected from land or seabed exposures, or from cored boreholes. Alternatively, a very useful estimate of rock density can be obtained from geophysical well logs. The gamma-gamma log, or as it is often referred to, the compensated density log, is a calibrated profile of bulk density. A schematic density log is shown in figure 7/26.

The principle of the method, its operation and interpretation of results are described in chapter 4.

If borehole data are sparse or not available, then seismic velocities derived from velocity analyses can be used as a means of determining density information. Such determinations are less precise than sample or borehole logging tests but do have the advantage that lateral variations in density within a large sedimentary basin can be accounted for. The basis of determining density values from seismic velocities is empirical and depends on experimental data relating p-wave velocity measurements in sedimentary rocks to laboratory measurements of saturated density. Figure 7/27

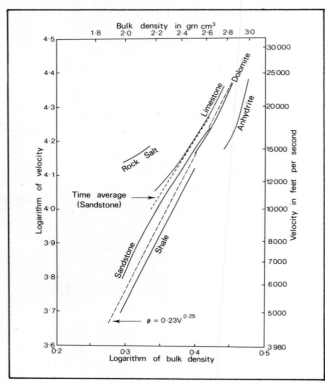

Figure 7/27: Relationship between saturated density of a variety of sedimentary rocks and their P-wave velocities as determined by laboratory measurements (after Gardner *et al*).

shows the relationship as determined by Gardner, Gardner and Gregory*.

It can be seen then that there is an interplay between gravity interpretation and seismic interpretation which can allow the seismic interpreter to utilise the results of seismic surveys to define the interpretation of gravity data, and by so doing, test the results of the original seismic interpretation. An example of the value of comparing seismic refraction and gravity survey data is described by Bacon and McQuillin. †

---

*G.H.F. Gardner, L.W. Gardner and A.R. Gregory, 'Formation velocity and density — the diagnostic basis for stratigraphic traps' *Geophysics* 39 (1974) p.776.
†M. Bacon and R. McQuillin, 'Refraction seismic surveys in the North Irish Sea' *Jl. Geo. Soc. Lond.* 128 (b) (1972) pp. 613-621.

References and suggested reading

M. Bacon and R. McQuillin 'Refraction seismic surveys in the North Irish Sea' *Jl. geol. Soc. Lond.,* 128, (1972) pp. 613-621.

G. Badtke, 'Better co-ordination of faults by combined reflection and refraction surveys' *Prakla-Seismos Report,* No. 1/76, pp. 10-11 (Prakla Seismos, Hannover, 1976).

G.H.F. Gardner, L.W. Gardner and A.R. Gregory 'Formation velocity and density — the diagnostic basis for stratigraphic traps' *Geophysics,* 39 (1974), pp. 770—80.

F.S. Grant and G.F. West, *Interpretation theory in applied geophysics.* (McGraw-Hill, New York, 1965).

Hydrographer of the Navy, *Admiralty manual of hydrographic surveying.* Vols. I and II. (HMSO, London, 1965).

R. McQuillin and D.A. Ardus, *Exploring the geology of shelf seas.* (Graham & Trotman, London, 1977).

A.W. Musgrave, *Seismic refraction prospecting.* (Society of Exploration Geophysicists, Tulsa, Oklahoma, 1967).

L.L. Nettleton, *Geophysical prospecting for oil.* (McGraw-Hill, New York, 1940).

Elementary gravity and magnetics for geologists and seismologists. No. 1 (Society of Exploration Geophysicists, Tulsa, Oklahoma, 1971).

D.S. Parasnis, *Principles of applied geophysics* (Methuen and Co, London, 1962).

R.E. Sheriff, *Encyclopaedia dictionary of exploration geophysics.* (Society of Exploration Geophysicists, Tulsa, Oklahoma, 1973).

M.M. Slotnick, 'A graphical method for the interpreration of refraction progile data' *Geophysics,* 15 (1950) pp. 163-80.

D.K. Smythe and N. Kenolty, 'Tertiary sediments in the Sea of the Hebrides' *Jl. geol. Soc. Lond.,* 131 (1975) pp. 227-233.

W.M. Telford, L.P. Geldart, R.E. Sheriff and D.A. Keys *Applied geophysics* (Cambridge University Press, Cambridge, 1976).

H.R. Thornburgh, 'Wavefront diagrams in seismic interpretation *Bull. AAPG,* 14 (1930) pp. 185-200.

8. HYDROCARBON RESERVOIRS AND THEIR DETECTION

Although the seismic interpreter is involved at times in work associated with exploration for resources other than hydrocarbons, coal for example, by far the most important application of reflection seismics is in the oil industry. Here the search is for geological structures which are prospective in that there is a reasonable probability that they form traps enclosing exploitable quantities of oil or natural gas. Such structures vary widely in type but always have certain general characteristics; the essential properties of an oil or gas reservoir. As well as describing the nature and classification of reservoirs, this chapter also describes the recently developed methods for detecting hydrocarbons within them.

The interpreter's role extends beyond the initial exploration for prospective structures in advance of drilling. Following the discovery of oil or gas in a structure, an important task is that of mapping in great detail the size of the reservoir to assess its commercial viability. A following chapter on the Kingfish Oil Field provides a good illustration of the importance of reliable and accurate seismic mapping to proper planning of the development of an oilfield. Mapping of seismic events is the interpreter's main task; however, study of seismic sections can give crucial information on changes in lithology within rock groups, changes which can be of paramount importance to the development of exploitable resources. The case history of the Rainbow Lake discovery (chapter 10) shows how such lithological variations can be detected on seismic sections and related to the prospectivity of map structures.

8.1 The formation of hydrocarbon reservoirs.

Hydrocarbon reservoirs are found within sedimentary basins in which rock sequences have accumulated to a sufficient thickness to allow diagenetic maturation of those rocks which are the original source of oil and gas, the source rocks. Maturation of the organic material in the source rocks demands a high enough temperature (possibly between 100 – 200 °C) for certain chemical changes to take place which are essential to the production of petroleum fluids. In addition, a high overburden pressure is needed to initiate the process of expulsion of these fluids from the source rocks. The fluids then pass through various stages of migration until eventual entrapment of the oil and/or gas as pools in hydrocarbon reservoirs.

Thus in exploration it is important to establish in the first place whether conditions within a sedimentary basin were likely to have been capable, at some time during its history, of producing sufficient organic source material as well as the correct environment for primary production and migration of hydrocarbon fluids. Secondly, it is necessary to establish whether suitable source and reservoir rocks actually occur within the rock sequence, or in the absence of definitive borehole data, to assess the likelihood of such

occurrences from an evaluation of palaeoenvironmental considerations. Thirdly, the structural history of the basin is studied to assess whether or not any phase of excessive tectonic activity might have resulted in the loss of previously formed hydrocarbons. Finally, the relationship is assessed between present structural features and those associated with earlier phases of basin development, thus hopefully establishing an integrated history of basin development, hydrocarbon generation, tectonic activity and hydrocarbon migration. The principal elements of the hydrocarbon-forming process can be summarised as follows:

1. Development of a sedimentary basin.

2. Deposition of source rock material: normally organic clays deposited in an anaerobic environment, or in some cases, peaty formations which will form residual coal deposits as well as providing source material for natural gas and/or petroleum.

3. Deposition of porous and permeable reservoir rocks, usually either sandstones or carbonates. Some carbonate reservoirs owe their porosity and permeability to diagenetic and tectonic modification of original sediments. It is important that reservoir rocks are interbedded with potential cap rocks.

4. Burial and diagenesis of source and reservoir rocks leading to formation of hydrocarbons and primary migration of fluids into porous rocks; this primary migration may take place at least partly in aqueous solutions.

5. Tectonic modification of the basin to form a variety of trap structures.

6. Secondary migration of mature hydrocarbons as a result of tectonic activity and increases in pressure due to increased depth of burial. All petroleum fluids have a lower specific gravity than water and migrate upwards through the influence of gravity; however, hydrodynamic and structural effects may induce considerable lateral migration, and in some cases even downward migration.

7. Capture of migrating fluids in hydrocarbon traps consisting of porous and permeable reservoir rock capped by impermeable strata which seal the structure and inhibit further migration thus leading to development of oil/gas pools in the reservoir.

It should be noted that the above summary is very simplified; processes associated with the origin and migration of petroleum fluids are not fully understood and are still the subject of some controversy. The history of a real sedimentary basin is usually complex including numerous phases of marine and non-marine deposition associated with cycles of marine transgression and regression; pulses of tectonic activity may occur at different times during its development leading to major unconformities as well as dislocation and folding of the strata, whereas other structures may have greater continuity of expression such as major growth faults which often control sedimentation across horst blocks and at basin margins.

8.2 Hydrocarbon traps

Traps can be broadly classified as of three different types; structural traps, stratigraphic traps and combination traps (partly structural and partly stratigraphic). Not every hydrocarbon-bearing structure fits neatly into this classification but the exceptional types are of relatively minor economic importance. In every hydrocarbon trap the hydrocarbons are contained within a body of reservoir rock which is porous and permeable and escape from this body of reservoir rock is prevented, both upwards and laterally, by enclosure of the reservoir in impermeable strata, sometimes called the cap rock. If both oil and gas are present, gas will occupy pore space in the reservoir rock in the highest part of the trap. Below this will be a layer of oil and below the oil layer pore space will be water-filled. The contact between gas and oil is termed the gas/oil

contact or GOC, that between oil and water the oil/water contact or OWC. If only gas occurs, the only contact will be between gas and water, a gas/water contact or GWC. These contacts are usually horizontal or very nearly horizontal. In some circumstances water flow in the reservoir or the effects of varying capillarity, can render this contact substantially non-horizontal, but such circumstances are relatively uncommon.

Structural traps are formed by post-depositional deformation of rock strata with development of folds, domes, faults and unconformities. Stratigraphic traps are formed as a result of the occurrence of lithological variation in strata which is directly related to environmental conditions at the time of deposition. Lenticular sand bodies and reef structures are among the more important stratigraphic traps. Stratigraphic traps are generally more difficult to detect by the seismic method than structural traps. Within an oil province, exploration for such traps is often of increasing importance as time passes and all the obvious structural traps have been drilled. Combination traps include an element of both stratigraphic and structural control and within this group traps associated with unconformities and with salt movements are among the more important.

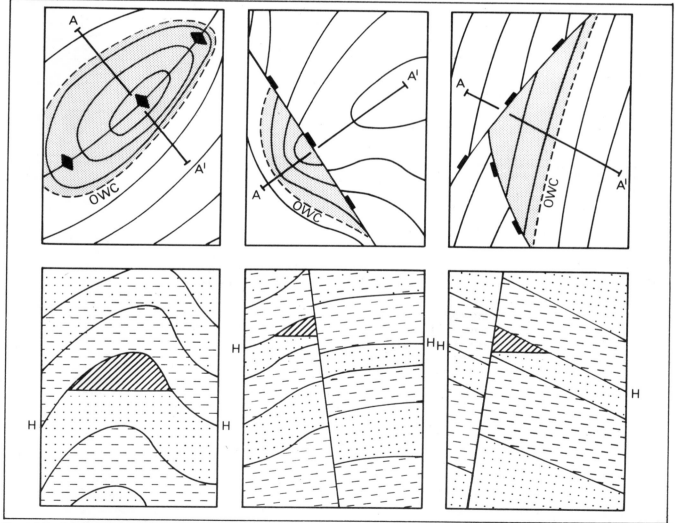

Figure 8/1: Schematic structural traps showing (above) contours on a horizon H with geological sections A—A' (below). On the left a structure over a closed dome; in the centre, a faulted anticline with closure; on the right, a fault-controlled closure in beds having homoclinal dip.

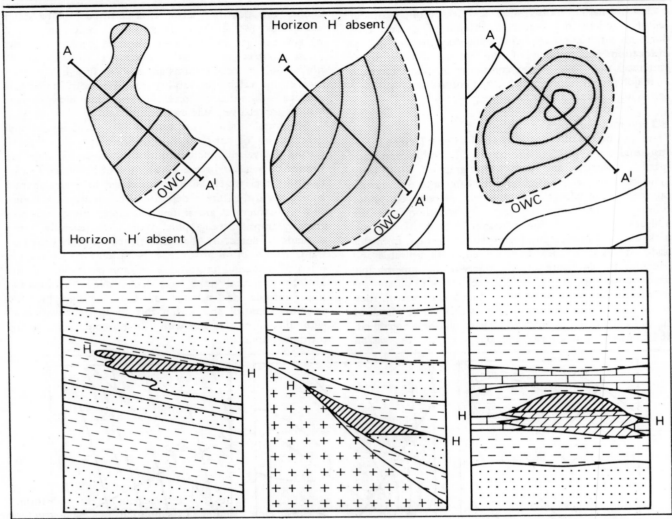

Figure 8/2: Schematic stratigraphical traps showing (above) contours on horizon H with geological sections A—A' (below). On the left a lenticular body of sand; in the centre, a pinchout closing against a basement high; on the right, a reef.

Potential structural traps can be located on seismic horizon maps. They are characterised by closed structures as mapped on, or near, the top of a reservoir rock formation. For the trap to contain oil or gas it is necessary for the closed structure to be sealed by a suitable cap rock. Figure 8/1 is a schematic diagram showing typical structural traps as mapped on two-way time or depth-converted horizon maps, and in the lower part of the illustration, as seen on geological sections through each structure. Seismic sections along A—A' would appear similar to the schematic geological sections but without normally giving any direct indication of the oil/water contact. An important characteristic of structural traps is that they are often capable of providing oil and/or gas production from a number of horizons. If the rock succession contains a number of potential reservoir intervals, more than one of these might be sealed and thus a number of separate pools can be formed greatly enhancing the value of the field. Such a characteristic is particularly true of closed domes, but can also be the case with other types of structural traps.

To locate stratigraphic traps from seismic data, the quality and resolution required is usually greater than that required for detection of structural traps. A further problem is that of identifying the reservoir interval. With structural traps, a borehole sited on a structural high will usually penetrate a thick succession of beds, all of which have closure under the one well site. With stratigraphic traps, such as pinchouts for example, different prospective intervals will lap against older formations to give potential traps at locations laterally well separated. With such reservoirs as sand lenses, channel sands etc, there may be very little seismic evidence at all for the existence of a trap. Reef carbonates are exceptional in that they are subject to direct location by the seismic method; reefs form identifiable highs on seismic maps; also, the reef carbonates usually exhibit a high velocity compared with adjacent and overlying sediments.

Figure 8/2 shows schematically three important types of stratigraphical traps. The method of location of the sand lens would be to identify the fact that the reflector H was not continuous throughout the prospect and that it had an areal form akin to that of a known type of sand body such as a sand bar or buried channel. A map of the occurrence of H combined with an isochron map of the top of the body, or of a stronger reflector immediately above H could then be used to locate an exploration well. Figures 11/5 and 11/16

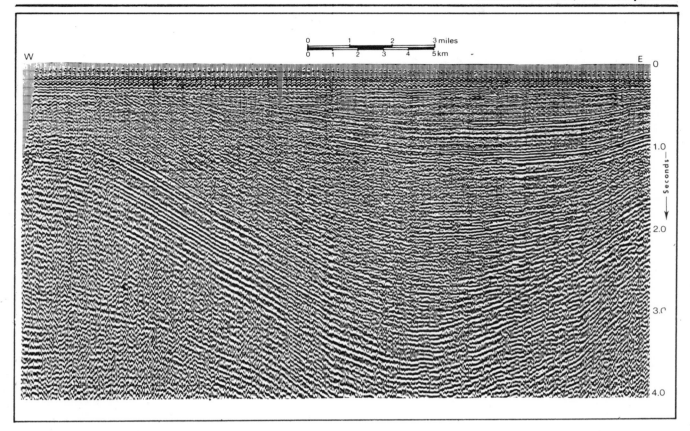

Figure 8/3: Seismic section from the Cardigan Bay area of the Irish Sea illustrating pinchout of bedding against a structural high.

(pp.164 and 173) show seismic sections across a well developed Miocene channel system. In this case the oilfield under discussion occurs at a deeper level but the ability to map channels using seismics is nevertheless well illustrated.

Pinchouts, such as illustrated in figure 8/2, are often detectable on seismic records. Figure 8/3 shows a seismic section from the southern Irish Sea and a number of intervals can be seen to thin out against a structural high to the east. The main problem is one of resolution and actual definition of the feather edge. A further complication can be that lateral sealing of such a reservoir can be due to lithological change, not closure.

The schematic reef structure shown in figure 8/2 is a simplified version of the type of structure described in chapter 10. A good example of a seismic section across a productive reef is shown in figure 10/7 (p.151).

In figure 8/4 are shown a number of combination traps. The first is a typical unconformity trap in which a productive sandstone reservoir is overlain by impermeable clays and sealed by younger impermeable clays above the unconformity. Closure depends on the one hand on a plunging unconformity and on the other a fault which seals off the highest part of the closure. This type of trap can be directly detected by seismic mapping. The second example shows a permeable reservoir sealed up-dip by change of lithology to impermeable rock, combined structurally with a plunging anticline to give an arcuate oil/water contact. Detection of this type of reservoir depends on an

ability to predict lithological change in the reservoir formation. The final example is a simplified salt dome structure with which is associated productive reservoir sands at a number of levels. The contour map shows the form of horizon H which is subject to complex doming and subsidence with attendant faulting leading to the development of a number of disconnected pools. Mapping complex structures around salt domes is a complicated task requiring high quality seismic data; nevertheless, seismic mapping can identify closures on which to site exploration wells.

8.3 Seismic stratigraphy
From the previous sections, the importance is evident of any information on lithology which can be derived from seismics. There are two lines of approach to this problem. The first is through an analysis of quantitative data on velocities, amplitudes, reflection polarities and frequencies, and recent advances in processing techniques have greatly improved the reliability of such data. The second approach is through identification on the seismic sections of reflection patterns which can be interpreted in terms of depositional features. Any final conclusion should be based on a combination of both lines of study and as in most interpretive work the reliability will be greatly enhanced if results can be integrated with and checked against well data.

Interval velocity can be calculated for an identified rock interval using stacking velocities derived from analyses made during processing to facilitate CDP stacking. The average velocity values for reflection events at the top and

117

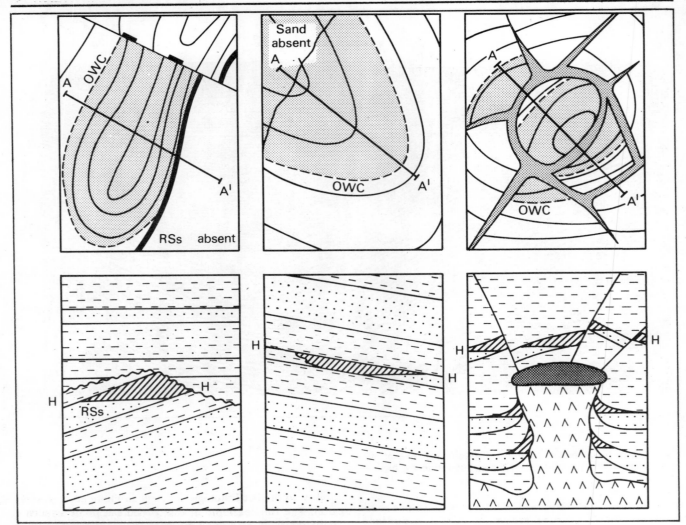

Figure 8/4: Schematic combination traps showing (above) contours on horizon H with geological sections A—A' (below). On the left an unconformity trap; in the centre a plunging anticline sealed up-dip by change in lithology; on the right a complex salt-dome structure.

base of an interval are used. The Dix formula can then be applied, but it should be emphasised that this formula only applies to a situation of flat lying, constant thickness bedding at the velocity analysis site. The Dix formula is given by:

$$V = \left[\frac{t_B \, \overline{V}_B^2 - t_T \, \overline{V}_T^2}{t_B - t_T}\right]^{\frac{1}{2}}$$

where V is the internal velocity, V_T and V_B are the rms velocities to the top and base of the interval and t_T and t_B are the reflection times.

In table 8.1 are listed P-wave velocity ranges for the main types of sedimentary rocks.

For most rock types a fairly wide velocity range applies and the principal factor affecting velocity is porosity. In a sedimentary basin, porosity of clastic rock tends to decrease with depth of burial due to compaction under over-burden pressure; thus it is often difficult to differentiate between, for example, sandstones and shales on the basis of interval velocity. Limestones and dolomites can sometimes be identified by having an anomalously high

Table 8.1: P-wave seismic velocities

Rock type	Velocity in m/sec
Unconsolidated sands, clays etc	600—2,000
Shales	2,000—3,500
Sandstones	2,000—5,000
Rock salt	4,500—6,000
Limestones	3,500—6,500
Dolomites	5,000—6,500

velocity compared with the surrounding velocities. Similarly, rock salt (and other evaporites) are usually identifiable by their velocity being higher within Mesozoic rocks and lower in Palaeozoics. If a particular rock interval exhibits, within a prospect, lateral variation of velocity, this can be due either to change of lithology, in particular variations in cementation and detrital mineralogy, or to tectonic effects. In the latter case, the post-depositional history of the basin is likely to have been such that the interval has been subject to deeper burial in some places than in others, thus affecting porosity and velocity. If well data are available, such lateral velocity variations can be compared with log and sample data and the cause of variation identified. Seismic data can then be used to interpolate lithological variations between wells. A particular application can be where a reservoir rock exhibits lithological variations which control its oil or gas production properties.

As is described in chapter 1 (p.3) the amplitude of a seismic reflection is determined by the reflection coefficient at the reflecting boundary. Conventional processing methods do not display reflections in terms of real amplitudes but acquisition and processing systems can be designed to retain these and a study of amplitudes can give useful diagnostic information. If the density variation across a reflector is not large, then the reflection coefficient is approximately given by the velocity difference across the interface divided by the sum of velocities. If the reflection takes place at an interface between high velocity rocks overlying low velocity rocks then the coefficient is negative and the reflection undergoes a change of polarity at the boundary. Such polarity reversals can be detected and for some applications colour presentations are used as a means of identification. The main application of real amplitude processing is in detection of hydrocarbons; however, in conjunction with studies of interval velocities reflection amplitude variation can be used to map lateral lithological change.

A full description of the range of sedimentary features detectable on seismic sections and their significance in terms of the historical development of a sedimentary basin is too large a subject for full treatment here (see list of suggested reading on this subject). Some features are described in the case histories (chapters 9-11); reefs, channels and growth structures across faults and horsts. Marine transgressions and regressions can be recognised through identification of onlap of sediments associated with an advancing shoreline and offlap as associated with a retreating shoreline. Deltaic fans can be identified and mapped and deductions can be made regarding direction of sediment transport during deposition of such formations. Formations which show large lateral variation in thickness are more likely to have been deposited in shallow water whereas formations which exhibit uniform bed thicknesses over large areas are more likely to be deep water sediments. Post-depositional deformation due to flowage can indicate the presence of either salt or mobile clays. Deposits above the mobile layer will exhibit folding which is not present in strata beneath the mobile layer. The effect of plastic flow may be apparent also in controlling growth structures in horizons which were deposited during the period of flow. Velocity data will allow differentiation between salt and clay at the level of flow.

8.4 Detection of hydrocarbons

The presence of gas in a reservoir rock has often the property of producing a significant reduction in seismic velocity. The presence of oil does not have such a significant effect. In some circumstances it is even possible to see on the seismic section the actual gas/oil or gas/water contact, see figure 8/5. More usually the only detectable effect is the production of amplitude anomalies, these being due to the large negative reflection coefficient at the interface between gas filled reservoirs and the overlying cap rock. Such amplitude anomalies are often termed 'bright spots'. Not all amplitude anomalies are caused by the

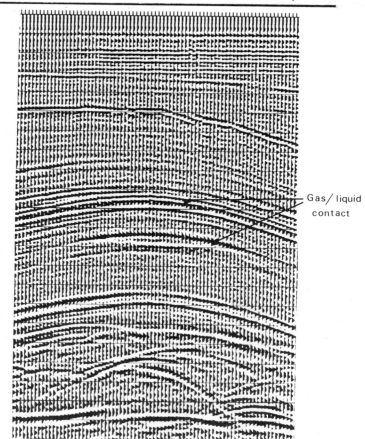

Figure 8/5: Seismic section showing gas/liquid contact in reservoirs.

presence of gas; therefore, it is important to interpret such anomalies with care and to take into account other possible hydrocarbon indicators such as phase change of reflection and pull-down of events beneath the prospective zone, as well as judging the geological feasibility of the anomaly being associated with a gas pool. To facilitate the study of amplitude anomalies, special processing procedures are adopted. The conventional seismic section has undergone a process whereby amplitudes throughout the depth range are subject to either automatic gain control or some form of trace equalisation. This process allows clear presentation of low energy reflections alongside high amplitude events but gives a distorted view of the true relative amplitudes. Modern digital acquisition systems allow recovery of data on true amplitude and the only problem is that of devising a suitable display mode. If normal display modes are adopted, then the true amplitude section will show little information other than the very high amplitude events. One useful alternative is a colour presentation and in this case it is possible to display data so that both amplitude and polarity are easily identified. Thereby it can be seen whether the peak or trough of a reflection is the larger and perhaps more importantly, whether the reflection has occurred at a boundary with positive or negative reflection coefficients (see figure 8/6, p. 133). An alternative approach is to employ an automated technique specifically designed for hydrocarbon detection. One such technique is the Hi-Scan system of Petty-Ray Geophysical of which an example is shown in figure 8/7. This system allows an initial selec-

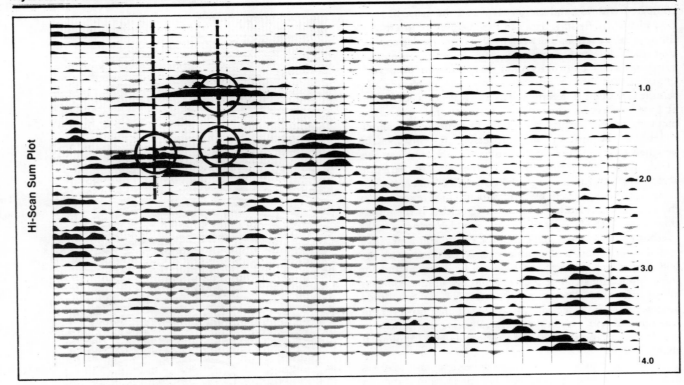

Figure 8/7: Automated hydrocarbon detection, Hi-Scan method. (*Courtesy: Petty-Ray Geophysical*).

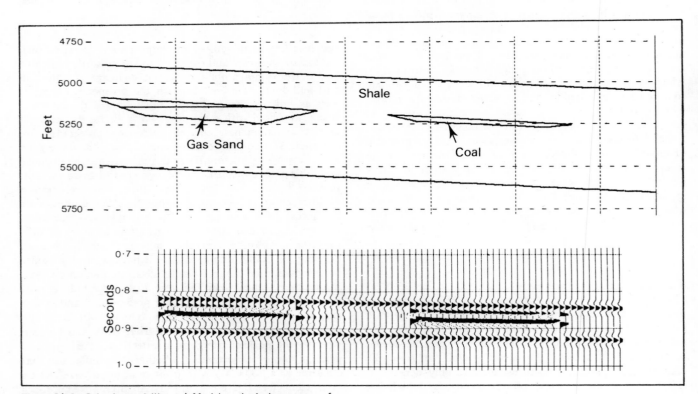

Figure 8/10: Seismic modelling. **a)** Model geological structure of a gas sand (left) and a coal seam (right) enclosed in a shale unit. **b)** A synthetic section showing the similarity between the effects of the two geological models. (*Courtesy: GeoQuest International Ltd/SCC*).

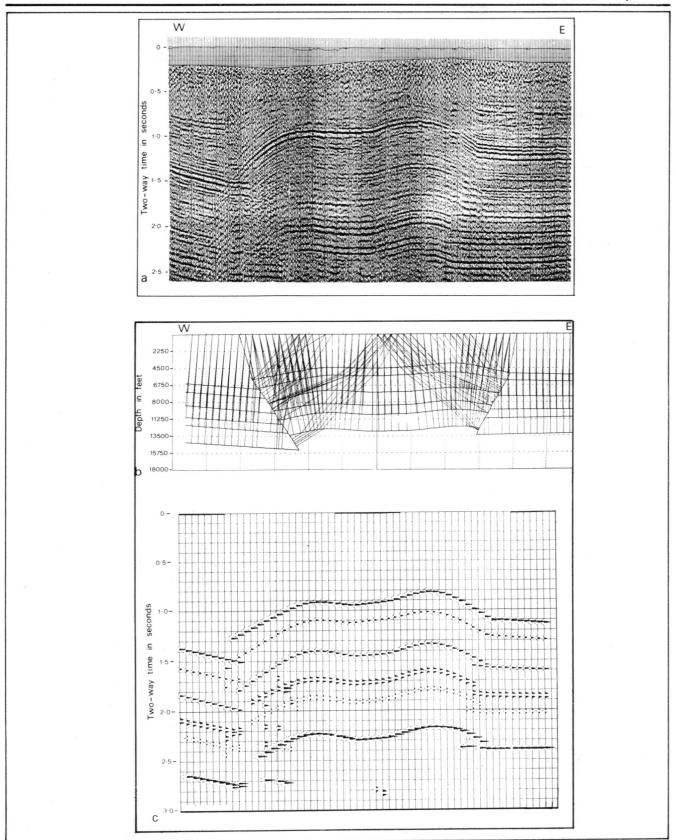

Figure 8/9: Ray-path seismic computer modelling. **a)** Field section from Wyoming showing faulting and diffractions. **b)** Geological model with computed ray paths, including diffraction paths. **c)** Synthetic seismic section which gives good fit to original section. The good fit between original and synthetic sections confirms the valid-ity of the geological model of a reverse-thrusted anticlinal horst block.
(*Courtesy: GeoQuest International Ltd/SSC*).

tion of four out of eleven possible hydrocarbon indicators; the data are then automatically scanned horizontally to yield summation plots which highlight those zones where there is corroborative evidence of hydrocarbon presence.

A further aid to hydrocarbon detection through seismic processing, Seislog*, was described in chapter 3. The latest presentation method for this process now provides a colour display superimposed on the Seislog* pseudo-sonic traces with up to 30 velocity intervals, each represented by a different colour tone. A distinct colour is not provided for each lithology in that several types and different ages of rocks may have the same velocity. The colour section exhibited in figure 8/8 (p. 133) shows how the method was used successfully to extend the boundary of a known Upper Devonian carbonate gas field in western Canada. On this section the pseudo-sonic traces are velocity contoured and high velocity carbonates show up in the blue range of colours, darkest blues representing highest velocities. The Crossfield interval is one such high velocity interval. An established gas field existed in an area to the left of the section whose limit was placed at the thick pseudo-sonic trace. The Seislog* study indicated an extension of the porous productive zone (light blue colour)

towards the right of the section and this extension was subsequently proved by drilling.

An increasingly important aspect of seismic interpretation is the development of digital computer models of the field section anomalies, through the derivation of supposed geological analogues. As required, and where hydrocarbon anomalies are interpreted, porosity and fluid effects can also be input. Figure 8/9 shows one method, the Aims* modelling system (Geoquest International Ltd). A simulated seismic shooting programme is generated, taking into account all normally incident ray paths, obeying Snell's Law and taking into account all diffraction generation. The field seismic section, from the Wind River Basin, Wyoming, is successfully modelled as being an anticlinal horst block.

In figure 8/10, use of the same program is demonstrated in an interpretation involving the identification of gas sands within a sand/shale sequence. The model shows the effect of a 25ft coal seam to be hardly different from that of a 50—60ft gas and water-filled sand. Thus care must be taken in using such interpretation methods in that modelling techniques do not give unique solutions but only demonstrate the fit between actual observations and one or more of a range of geologically plausible models.

References and suggested reading

R.E. Chapman, *Petroleum geology. A concise study* (Elsevier, Amsterdam, 1973).

E.E. Cook and M.T. Taner, 'Velocity spectra and their use in stratigraphic and lithologic differentiation' *Geophys. Prosp.*, 17 (1969), pp.433-48.

M.B. Dobrin, 'Seismic exploration for stratigraphic traps' *AAPG Memoir 26*, pp. 329-352. (1977).

A.I. Levorsen, *Geology of petroleum.* (W.H. Freeman, San Francisco, 1959).

L.D. Meckel and A.K. Nath, 'Geologic considerations for stratigraphic modelling and interpretation.' *AAPG, Memoir 26,* pp. 417-438 (1977).

N.S. Neidall and E. Poggiagliolmi, 'Stratigraphic modelling and interpretation — geophysical principles and techniques.' *AAPG, Memoir 26,* pp. 389-416. (1977).

R.F. O'Doherty and N.A. Anstey 'Reflections on amplitudes.' *Geophys. Prosp.*, 19 (1971). pp. 430-58.

R.M. Pegrum, G. Rees and D. Naylor, *Geology of the northwest European Continental Shelf.* Vol. 2. (Graham & Trotman, London, 1975).

W.L. Russell, *Principles of petroleum geology.* (McGraw-Hill, New York, 1960).

M.W. Schramm, E.V. Dedman and J.P. Lindsey, 'Practical stratigraphic modelling and interpretation.' *AAPG Memoir 26,* pp. 477-502 (1977).

M.T. Taner and R.E. Sheriff 'Application of amplitude, frequency and other attributes to stratigraphic and hydrocarbon determination.' *AAPG Memoir 26*, pp. 301-328 (1977).

9. MORAY FIRTH CASE STUDY

In this chapter we describe the interpretation of a regional seismic survey in the inner part of the Moray Firth, Scotland. In this area both borehole information and surface geology maps provide geological control; valuable information can also be obtained from gravity and aeromagnetic maps.

9.1 Surface geology

On land, the Moray Firth is mostly bounded by thick sequences of Old Red Sandstone and older rocks, but outcrops of Mesozoic rocks occur in a narrow coastal strip; Permian to Jurassic rocks are found in the Lossiemouth area and Triassic to Jurassic rocks crop out between Golspie and Helmsdale and at Ethie and Port and Righ (figure 9/1).

The presence of these scattered Mesozoic outcrops gave rise to the hypothesis that an extensive Mesozoic basin was present offshore; shallow seismic (sparker) surveys and shallow boreholes confirm this idea. At the time the seismic survey to be described here was commissioned by the Institute of Geological Sciences, the state of knowledge of the surface geology was as shown in figure 9/1.

We shall see later that normal faulting controls the structure of this part of the Moray Firth. Since it is not easy to identify faults on shallow sparker sections, and since most of the faulting, as we shall see, does not affect the shallowest horizons, little was known of the fault patterns in the basin prior to the seismic survey. On land, the major fault of the area is the Great Glen Fault, a wide fault zone along which transcurrent movement is believed

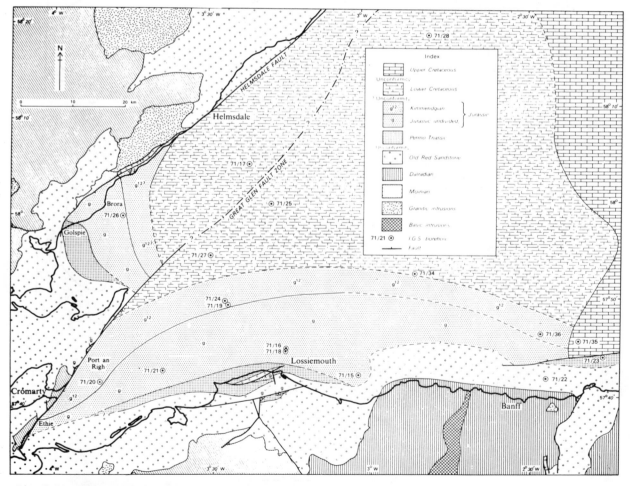

Figure 9/1: Solid geology of Moray Firth, as known before the seismic survey was commissioned (adapted from J.A. Chesher, et al. 'IGS marine drilling with m.v. *Whitethorn* in Scottish Waters 1970-71' *Inst. Geol. Sci. Rep.* no 72/10 (1972).

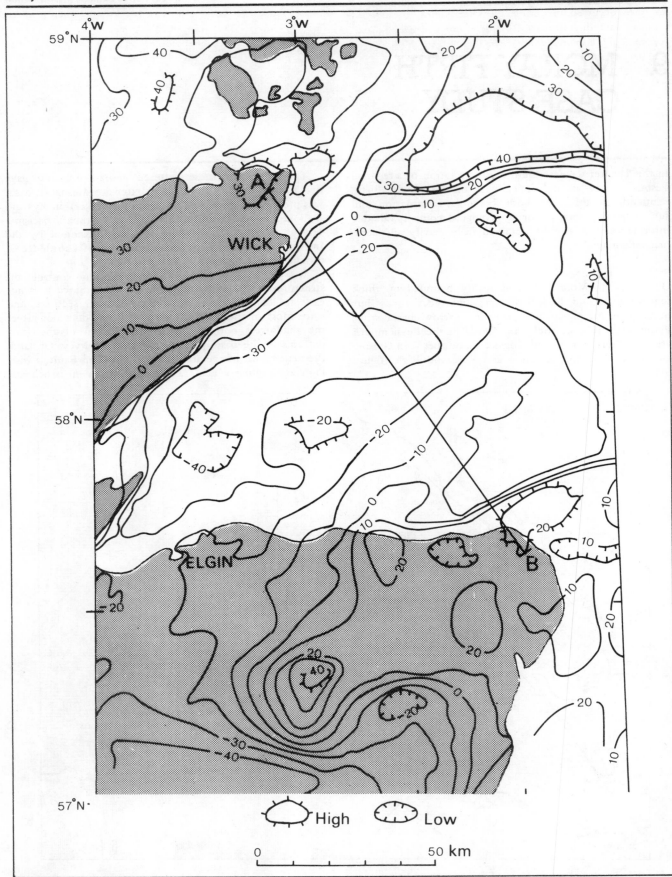

Figure 9/2: Bouguer anomaly gravity map of the Moray Firth.
(*Courtesy: IGS*).

to have occurred in Middle Old Red times, though the amount and direction of shift are controversial. A later, Tertiary phase of movement has also been postulated. This major fault zone was thought to continue in a north-north-easterly direction across the Moray Firth to link up with known faults on the Shetland Isles. The only other major fault known on land was the Helmsdale Fault running along the northeastern coast of the Moray Firth and down-faulting Mesozoic strata against Old Red Sandstone.

9.2 Gravity and aeromagnetic maps

The hypothesis of a deep Mesozoic basin in the Moray Firth received support from gravity and aeromagnetic data. There is an extensive gravity low in the Moray Firth (figure 9/2). When this was first discovered (with a very wide spacing between observations, resulting in a generalised version of figure 9/2), it was attributed to an extensive granitic body. Doubt was cast on this interpretation by the aeromagnetic survey (figure 9/3), which showed that over the gravity low the magnetic anomalies were of long wavelength and low amplitude, consistent with the presence of a thick sedimentary cover over magnetic basement.

The mapping of the surface geology described above confirmed the sedimentary basin interpretation. An approximate depth of the basin can be obtained by modelling a profile across it. Figure 9/4 shows such a model. First, a linear regional field was removed, based on the observed Bouguer anomalies at either end of the profile. A density contrast of 0.4g/cc was assumed between the Mesozoic rocks of the basin and the underlying Old Red Sandstone and the Dalradian, since published determinations of density from land outcrops in the area gave a mean value of 2.3g/cc for the Permian and Mesozoic rocks and 2.7g/cc for the Old Red Sandstone and Dalradian rocks. Exceptionally, the density contrast near A (figure 9/4) was taken to be only 0.3g/cc; in this region the Mesozoic sequence is not present and the density contrast is found within the Old Red Sandstone.

The two-dimensional model calculated to fit the observed anomalies then shows that along AB the depth of the basin does not exceed 4km, though owing to the uncertainties in regional field and density contrast this value is probably accurate only to about 0.5km. The basin is clearly fault-controlled, as indicated by the steep dip (50°) deduced for the southern margin in this model. Further steep gradients indicate major faults on the northern and northwestern margins of the basin, the north-western one being the seaward continuation of the Helmsdale fault line.

9.3 Seismic sections

To investigate the deep structure of the inner part of the Moray Firth, the seismic grid shown in figure 9/5 was shot. The survey is basically a 10 x 13km grid, with the more closely spaced lines intended to cross the Great Glen-Helmsdale Fault direction roughly at right angles. Geological identification of reflectors was initially based on correlation with outcrop data. However, lines 7 and 8 were planned to pass through the deep boreholes 12/21−1, 12/22−1 and 12/26−1. Line 8, through the first two of these, was of little use for identification because the Great Glen Fault passes too close to the tie to the rest of the survey. When information on borehole 12/26−1 was released, however, it showed that the identifications from

outcrop geology were fairly accurate (figure 9/6). Line 14 was shot after the interpretation of the rest of the data, when it became apparent that correlation across the Great Glen Fault (which runs roughly along line 10) was diffi-cult, so that the northern extremities of lines 1 to 6 were not clearly tied together; this line was positioned to run north of and roughly parallel to the fault in order to solve this problem.

Selected seismic sections are shown in figure 9/7, together with their interpretation. The events selected for mapping were as follows:

Horizon A: Intra Lower Cretaceous. This is a continuous shallow event in the eastern part of the area and is clearly defined except where it approaches seabed outcrop.

Horizon B: Base Lower Cretaceous. This is a very strong and continuous event especially in the eastern part of the area.

Horizon C: Originally described as mid-Jurassic, now known to be within Upper Jurassic. The event is not as strong as B or E but continu-ity is fairly good. The interval B−C is mainly lacking in countinuous reflection events.

Horizon E: Base Jurassic. (A horizon D lying just above, this event was considered for mapping but rejected because it showed very similar structural trends to E). This is the strongest deep reflector, with an easily recognised, well-defined character; it can be followed reliably over most of the area, although correlation across the Great Glen Fault is poor. There are many smaller faults which are very clearly defined at this level.

Horizon F: ?Intra Permo-Triassic. This deep event can be mapped only in the southeastern part of the area. Correlation across faults is poor and, although deep events are visible on some other lines, the uncertainty of correla-tion is too great to make possible any reliable extension of the mapping.

Identification of these horizons is illustrated in figures 9/6 and 9/8. To illustrate the difficulty of correlation across a major fault such as the Great Glen Fault, figure 9/7 (see pp.134-143) shows, for those parts of the line north of this fault, both the original picks based on seismic character and velocity information, and the revised picks after line 14 became available, enabling us to tie together all these line segments; in some cases drastic revision was required. Figure 9/8 shows how the outcrop geology known from IGS shallow boreholes fits into the structure of line 3. Once the identifications have been made, the seismic data suggests modifications to the outcrop geology map, of which the most conspicuous example is the southward extension of the lower Cretaceous outcrop as a result of the zone of faulting with downthrow to the southeast near line 12, which is an obvious feature of lines 4 and 5.

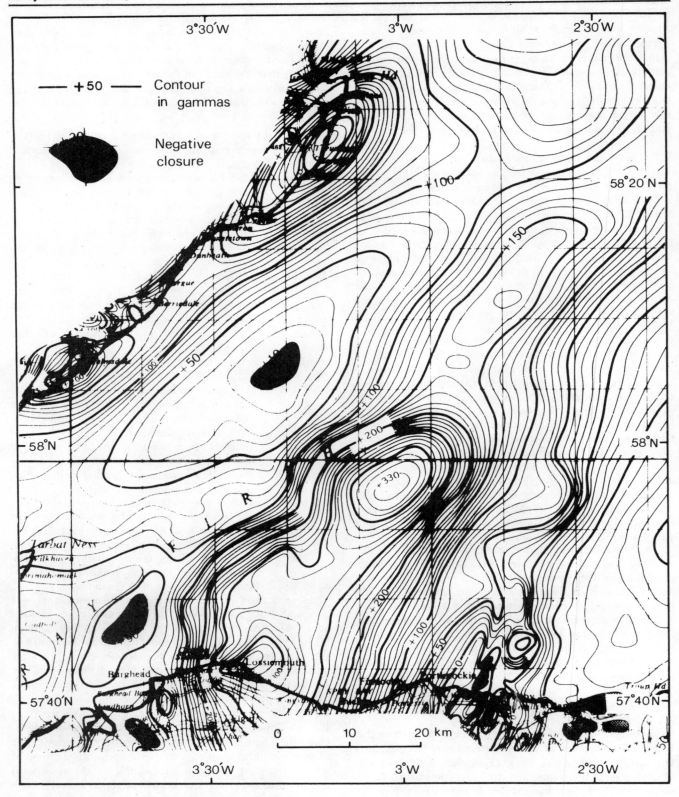

Figure 9/3: Aeromagnetic map of the Moray Firth area. (Adapted from IGS aeromagnetic map).

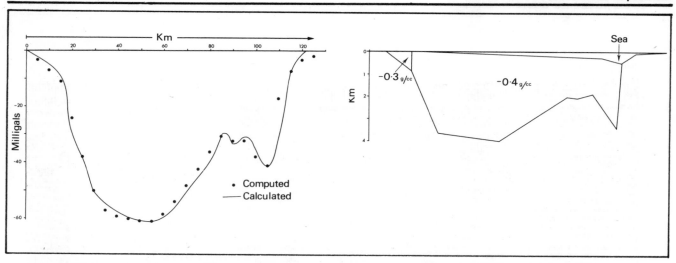

Figure 9/4: Interpretation of gravity profile. For location see figure 2. (After J. Sunderland. 'Deep sedimentary basin in the Moray Firth'. *Nature,* 236 (1972), pp.24—25).

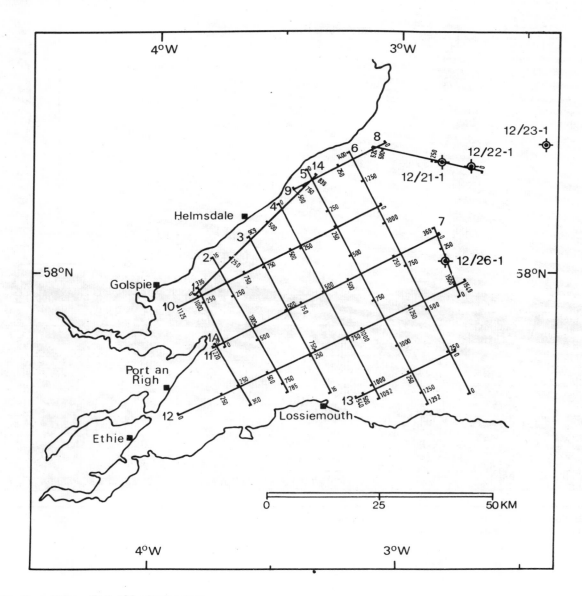

Figure 9/5: Plan of Moray Firth IGS seismic survey.

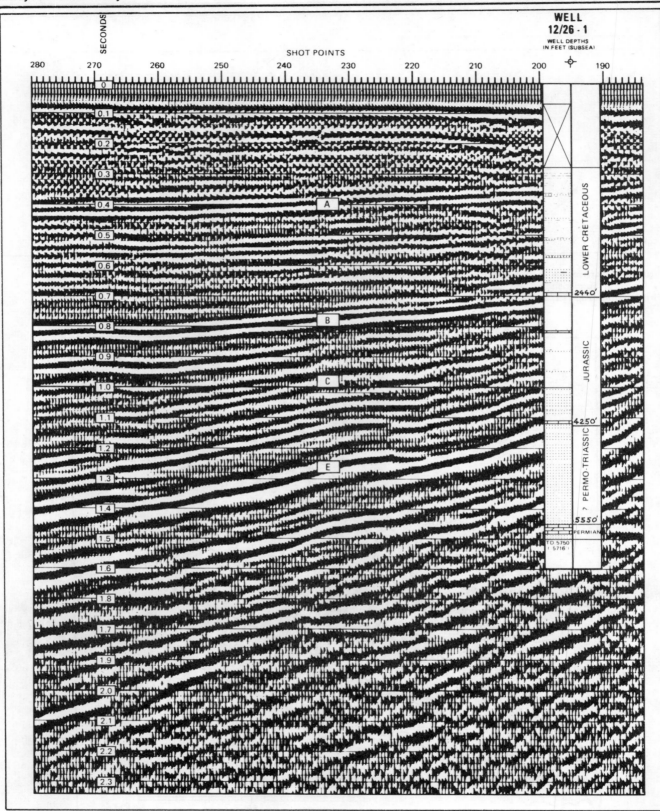

Figure 9/6: Correlation of seismic horizons and well stratigraphy.
(*Well section courtesy Hamilton, IGS seismic record, Seiscom survey*).

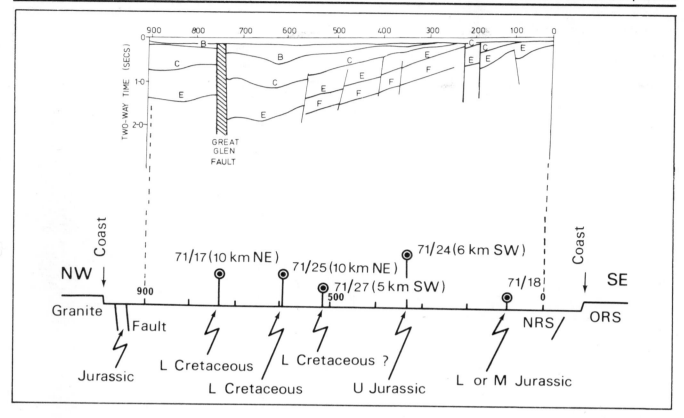

Figure 9/8: Correlation of line 3 with outcrop geology. IGS borehole identifications are shown.

Returning to figure 9/7, we can now comment on the details of the interpretation. Lines 3 to 5 are a series of cross-sections of the basin as it progressively deepens to the east. Line 3 shows the main basin structure bounded on the northwesterly side by the Great Glen Fault (hereafter abbreviated GGF). The southeasterly flank is broken by a number of small faults at the horizon E level, with downthrow towards the axis of the basin. At SP 100 and SP 200 there are faults which downthrow to the south east producing a shallow basin separate from the main basin. On line 4 the structure is similar, but the throw of the GGF appears to be very small; it is marked by a discontinuity at the deepest part of the basin. Again the south-easterly flank of the basin is broken by small faults giving rise to poor continuity of horizons C, E and F. At SP 775 and 820 there are major faults throwing down to the southeast. Across these latter faults correlation is not absolutely certain but is probably fairly reliable for Horizons B and E. On line 5 the GGF is again seen as a major fault, and reflector quality is poor on the upthrown side. Reflection quality of horizons B to E is good in the main basin. Major faults delineate a horst block between SP 660 and 770 with a sub-basin to the southeast of it.

Line 7, which has been included mainly to demonstrate the borehole tie, shows a fairly simple structure with all horizons rising southeasterly to a fault-bounded uplift zone. Line 11 shows the structure along the axis of the basin. There is a broad gentle arch in the southwest followed by a fairly regular dip to the northeast; however, much minor faulting is apparent.

9.4 Seismic maps

Two-way time maps for horizons A, B, C, E and F were constructed and are shown in figures 9/9 to 9/13. In general the connection of major faults from line to line was straightforward but the minor faults could not be treated so reliably and have simply been assigned trends similar to those of the major features. The increase in complexity of structure with depth is readily apparent. Horizon A (figure 9/9) shows a simple basin structure, faulted out against the GGF in the northeast and against a zone of uplift to the southeast; it may, however, be present again to the southeast of this zone, beyond the uplift zone at the end of line 7. At horizon B level (figure 9/10), the main basin and the southern sub-basin are already clearly defined, although only the major faults extend up to this level. A zone of uplift trending northeasterly through the main basin is also apparent. The deeper structure of Horizons C and E (figures 9/11 and 9/12) confirm these major structures, with the uplift zone in the main basin appearing as a horst block at the horizon E level. The fragmentary map of horizon F (figure 9/13) follows the same trends as horizon E. Putting together all the information derived from the seismic data, a structural summary map can be constructed, as shown in figure 9/14.

9.5 Conclusions

The IGS survey showed that there are some hydrocarbon prospects in the area. Not only is the Jurassic sequence, productive elsewhere in the North Sea, well developed, but there was a reasonable hope that more detailed work would reveal anticlinal closures especially on the axes of the horst which runs through the middle of the main basin. Since the IGS survey was completed a number of

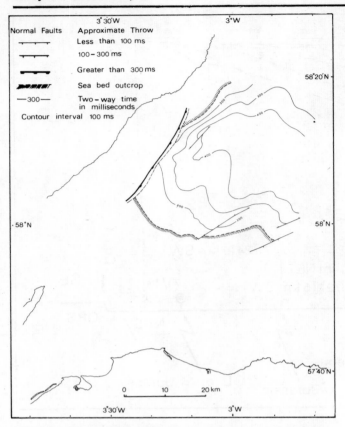

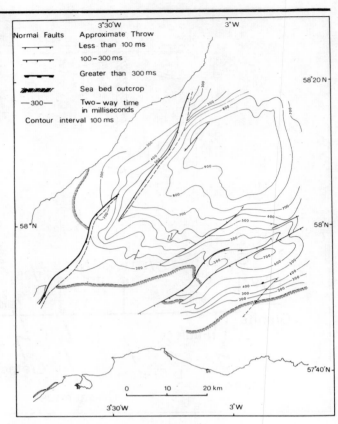

Figure 9/9: Two-way time map of horizon A. (All figures 9/9 to 9/14 are adapted from J.A. Chesher and M. Bacon. 'A deep seismic survey in the Moray Firth'. *Inst. Geol. Sci. Rep.* no. 75/11 (1975)).

Figure 9/10: Two-way time map of horizon B.

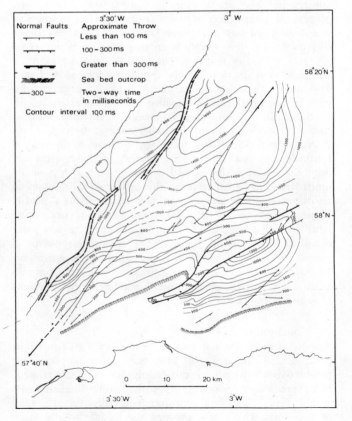

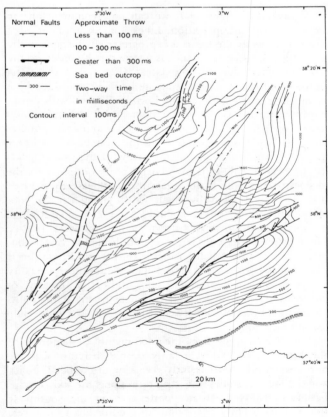

Figure 9/11: Two-way time map of horizon C.

Figure 9/12: Two-way time map of horizon E.

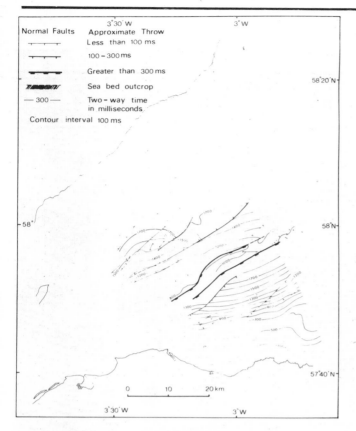

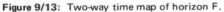

Figure 9/13: Two-way time map of horizon F.

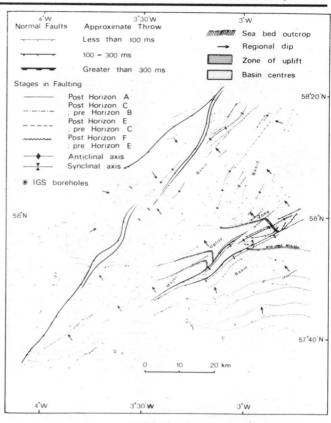

Figure 9/14: Structural summary map based on seismic information.

commercial surveys have been undertaken in the area and a much denser seismic coverage has become available to offshore operators. A series of wells (11/30–1 to 4) have delineated the Beatrice oilfield. This feature appears as a fault block on horizon E (figure 9/12) and a fault-bounded fold on horizon C (figure 9/11), but the IGS seismic grid spacing is too great to permit precise delineation of the trap.

Also, it should be noted that the IGS survey greatly increased knowledge of the general geology of the area.

Not only could a more precise outcrop geology map be prepared, but much information on the detailed pattern of faulting was obtained. In particular, the Great Glen Fault was traced across the area and shown to be sinuous and discontinuous. This pattern of outcrop makes it difficult to accept a Tertiary strike-slip movement along it, particularly when it is seen to be the largest of a series of faults in the area, all of which seem to be typical normal faults. Thus, a contribution has been made to a long-standing academic problem.

References and suggested reading

J.A. Chesher and M. Bacon, *A deep seismic survey in the Moray Firth* Inst. Geol. Sci. Rep. No. 75/11 (1975).

J.A. Chesher, C.E. Deegan, D.A. Ardus, P.E. Binns and N.G.T. Fannin, *IGS marine drilling with m.v. White-thorn in Scottish waters, 1970-71*, Inst. Geol. Sci. Report No. 72/10 (1972).

B.J. Collette, *Gravity expeditions 1948-1958*. Vol. 5, part 2. (Delft University Press, Delft, 1960).

G.Y. Craig, (ed.) *The geology of Scotland*. (Oliver & Boyd, Edinburgh, 1965).

D. Flinn, 'An interpretation of the aeromagnetic map of the continental shelf and Orkney, Shetland' *Geol. J.*, 6, (1969) pp. 279-292.

M.S. Garson and J. Plant 'Possible dextral movements on the Great Glen and Minch Faults in Scotland' *Nature Phys. Sci.*, 240 (1972), pp. 31-35.

R. McQuillin and M. Bacon, *Preliminary report on seismic reflection surveys in sea areas around Scotland, 1969-73*. Inst. Geol. Sci. Rep. No. 74/12 (1974).

J. Sunderland, 'Deep sedimentary basin in the Moray Firth' *Nature*, 236 (1972) pp. 24-25.

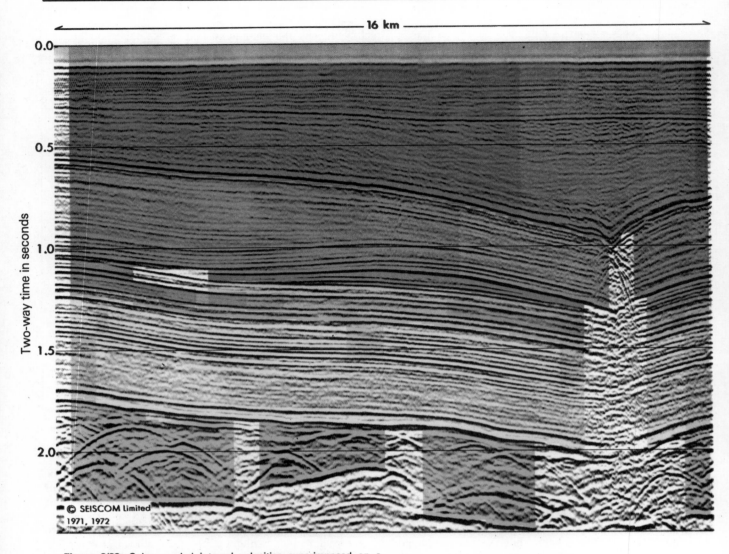

Figure 3/32: Colour coded interval velocities superimposed on a conventional North Sea section. Note the velocity gradient reversal at approximately 1.15s in the centre of the section. Colour code as follows: computed interval velocity in feet/second: pale blue — 5000; dark blue — 5,500—8,500; dark green — 9,000—10,500; pale green — 10,500—12,000; yellow — 12,500; orange — 13,000—15,000. (*Courtesy: Seiscom*).

Two-way time in seconds

Figure 8/6: Colour presentation of seismic section showing both amplitude and polarity of reflections. Amplitudes are scaled between 0 and 2, 2 and 6, and greater than 6. Amplitudes below 2 are suppressed. In the 2 to 6 range positive amplitudes are blue, negative black. Above 6 positive amplitudes are red, negative yellow. (*Courtesy: Prakla—Seismos*).

Figure 8/8: Seislog* colour presentation of section through gas field. (*Courtesy: Technika Resources Ltd*).

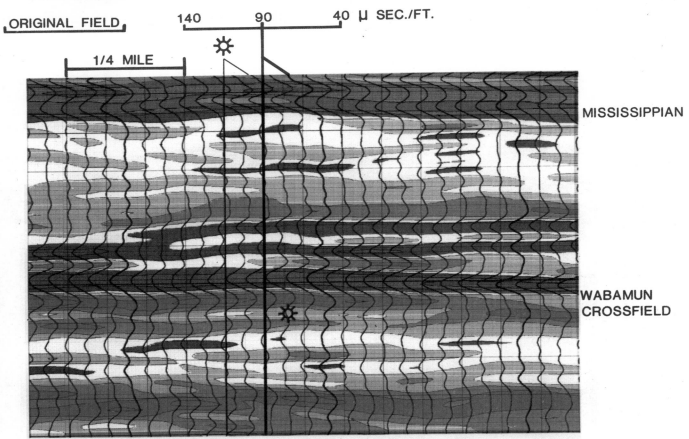

ORIGINAL FIELD

140   90   40 μ SEC./FT.

1/4 MILE

MISSISSIPPIAN

WABAMUN
CROSSFIELD

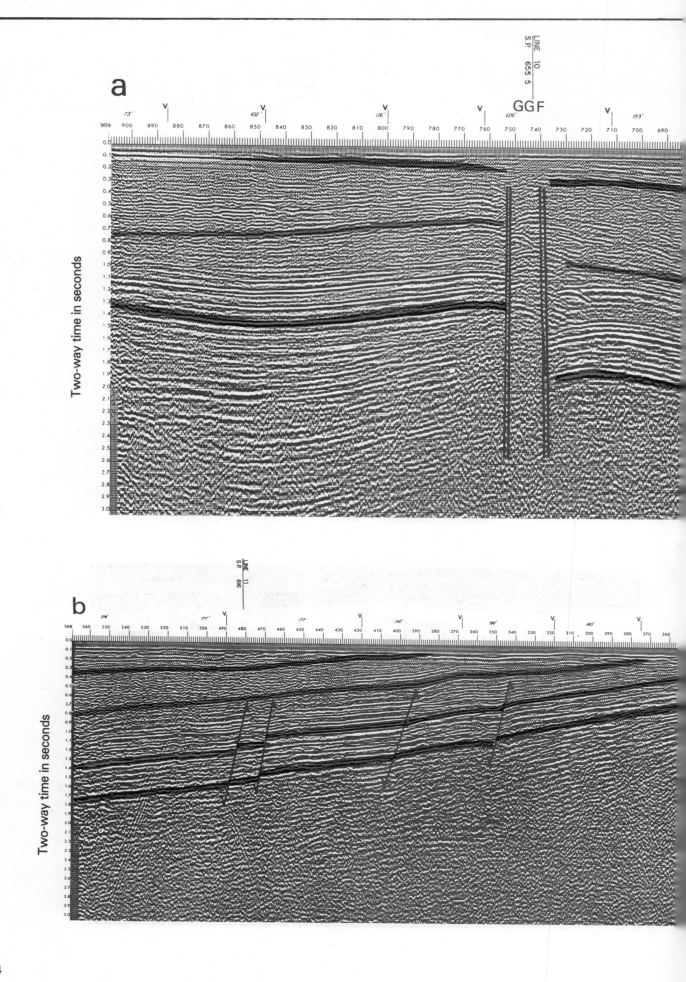

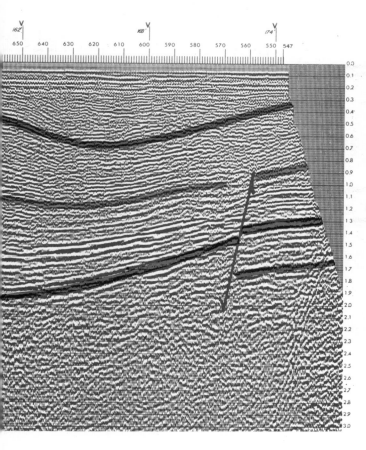

Figure 9/7: Selected seismic sections from Moray Firth survey. Letter V on top of record signifies location of a velocity analysis. Vertical scan is 0—3 seconds two-way time reflection time. Reflectors are colour coded: A — upper red; B — upper green; C — blue; E — lower red; F — lower green. Arrows on faults indicate direction of throw. On the northwestern parts of lines 4 and 5 the dashed coloured lines indicate horizon interpreted prior to survey of line 14. GGF marks the location of the Great Glen Fault zone. (*Courtesy: IGS records; Seiscom survey*).

Figure 9/7 a) Northwestern part of Line 3; SPs 909—547.

Figure 9/7 b) Southeastern part of Line 3; SPs 569—20.

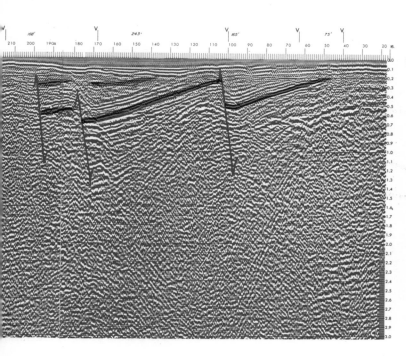

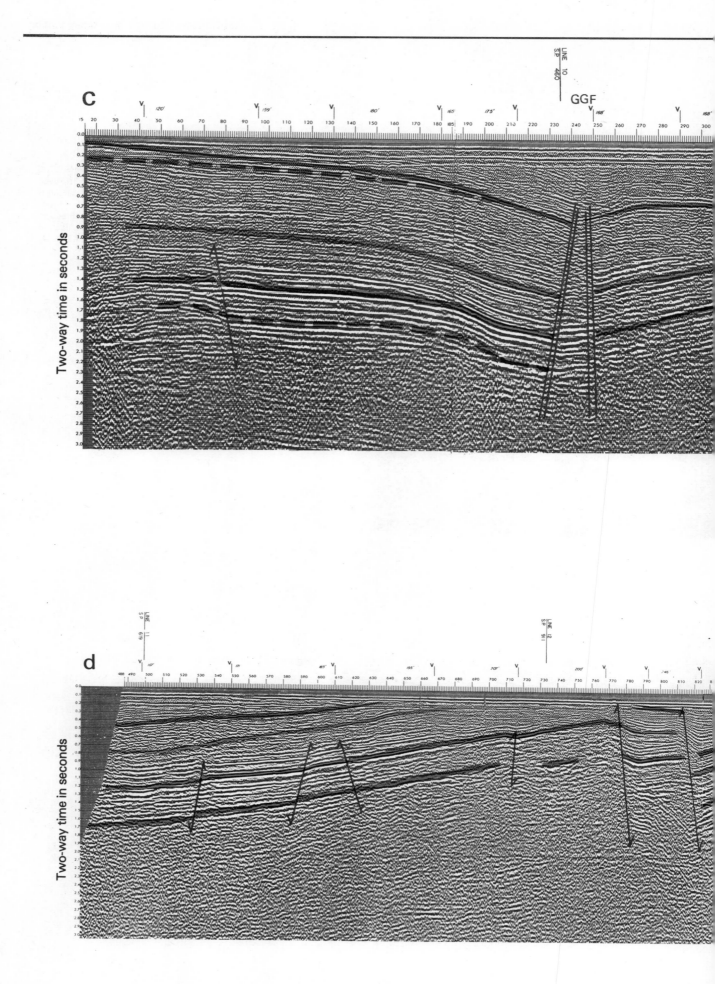

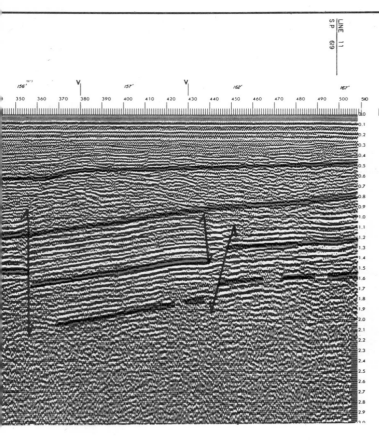

Figure 9/7 c) Northwestern part of Line 4; SPs 15—510.

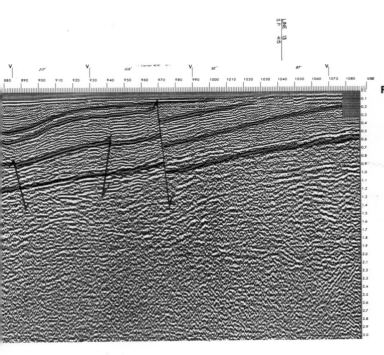

Figure 9/7 d) Southeastern part of Line 4; SPs 488—1092.

e

GGF

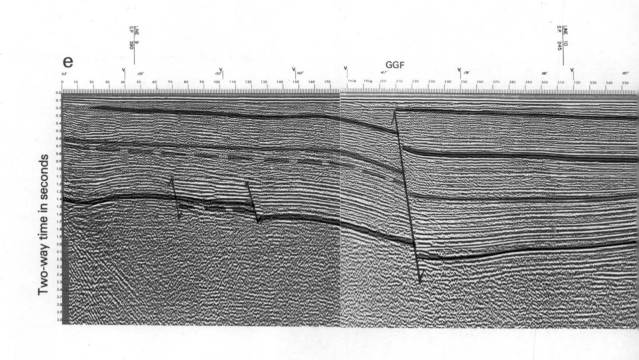

Two-way time in seconds

f

Two-way time in seconds

138

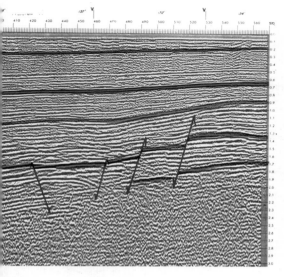

Figure 9/7 e) Northwestern part of Line 5; SPs 0—570.

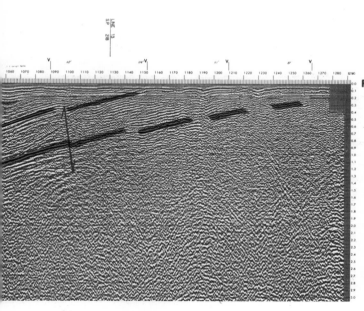

Figure 9/7 f) Southeastern part of Line 5; SPs 57—1290.

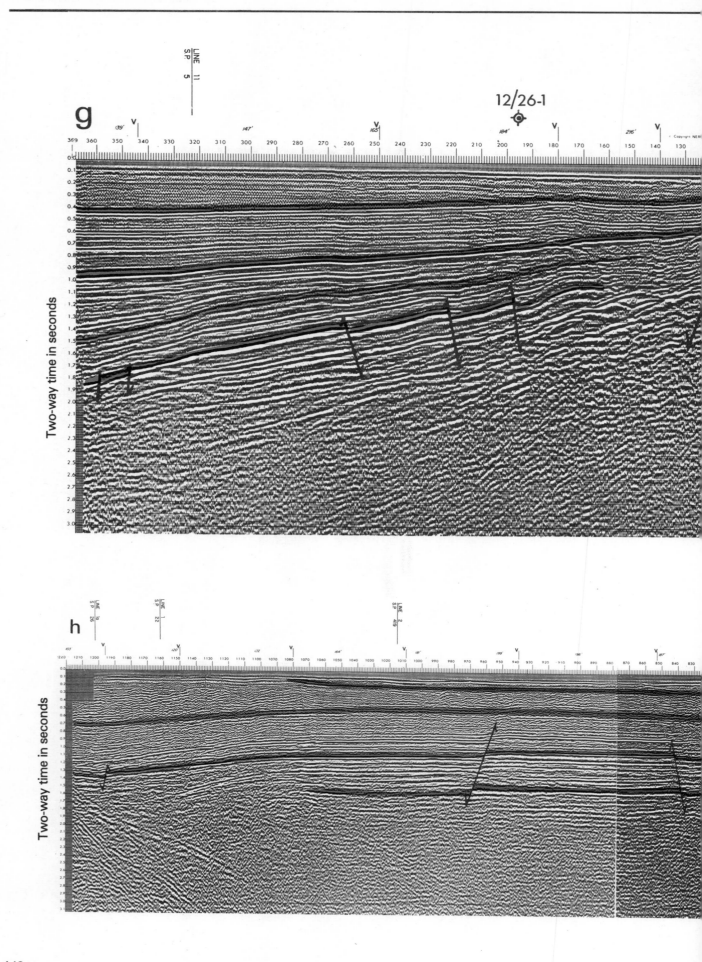

g

12/26-1

Two-way time in seconds

h

Two-way time in seconds

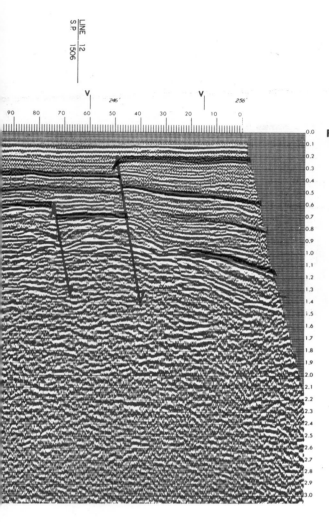

Figure 9/7 g) Line 7 showing location of well 12/26—1.

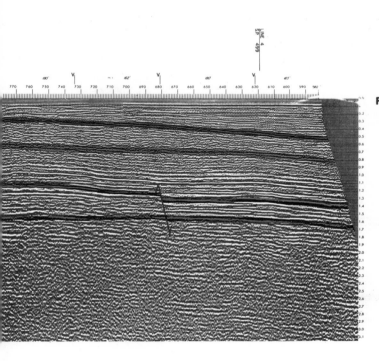

Figure 9/7 h) Southwestern part of Line 11; SPs 1220—582.

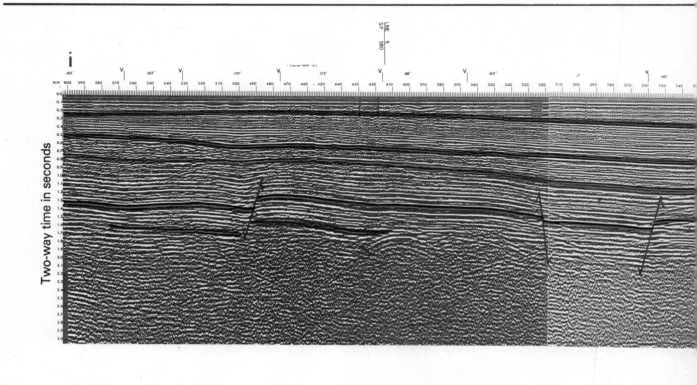

Figure 10/9: Frequency analysis of seismic data; window 0.8—1.05s. Note shift to high frequency content related to drape and thinning of beds over a reef buildup.
(*Courtesy: Aquitaine*).

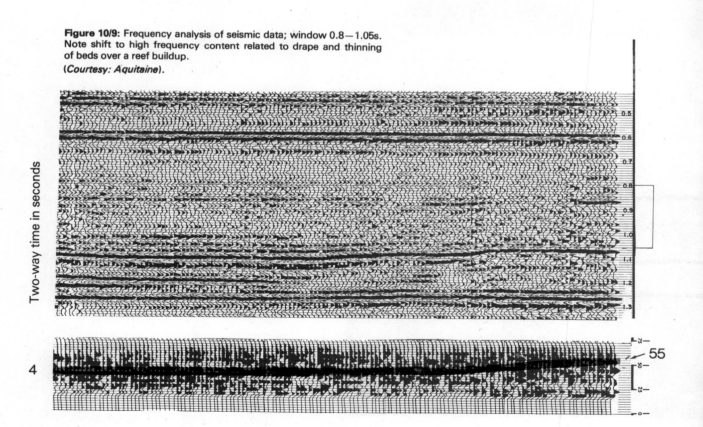

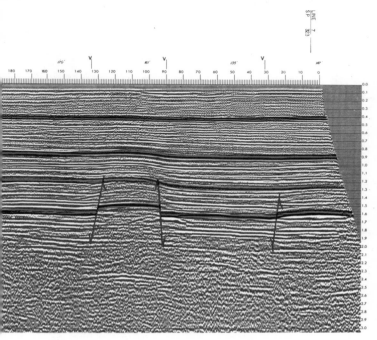

Figure 9/7 i) Northeastern part of Line 11; SPs 604—0.

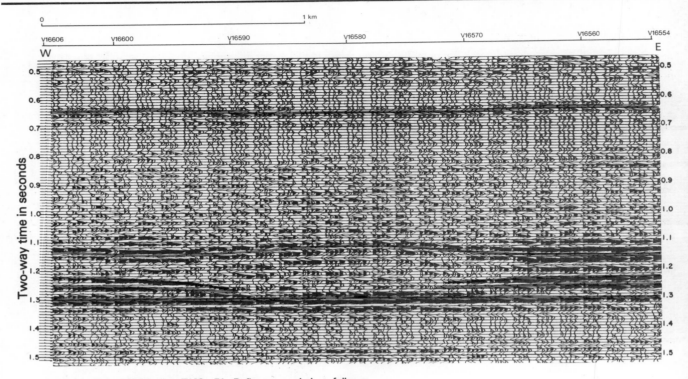

Figure 10/10: Seismic line T108—51. Reflectors coded as follows: top red — Wabamun; middle red — Slave Point; green — Top Black Creek Salt; bottom red — Base Cold Lake Salt. (*Courtesy: Aquitaine*).

Figure 10/11: Seismic line T108—44. (*Courtesy: Aquitaine*).

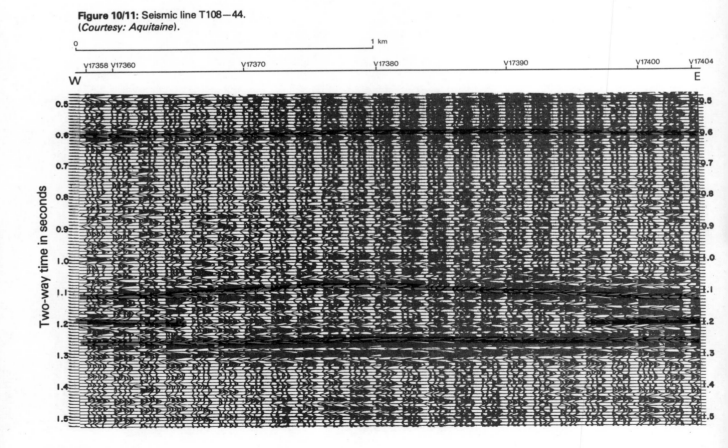

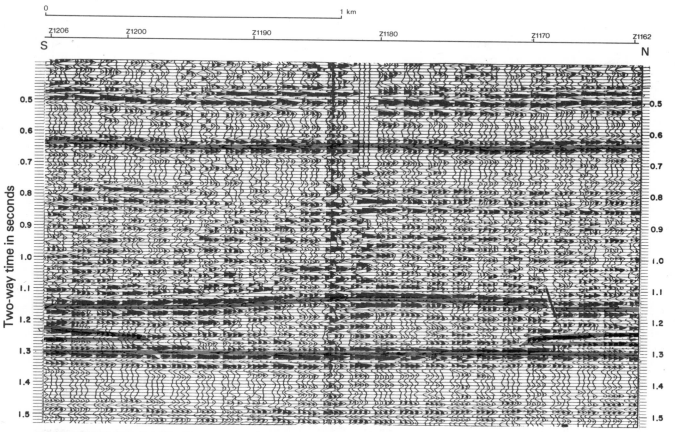

Figure 10/12: Seismic line R9—32.
(*Courtesy: Aquitaine*).

Figure 10/13: Seismic line R9—33.
(*Courtesy: Aquitaine*).

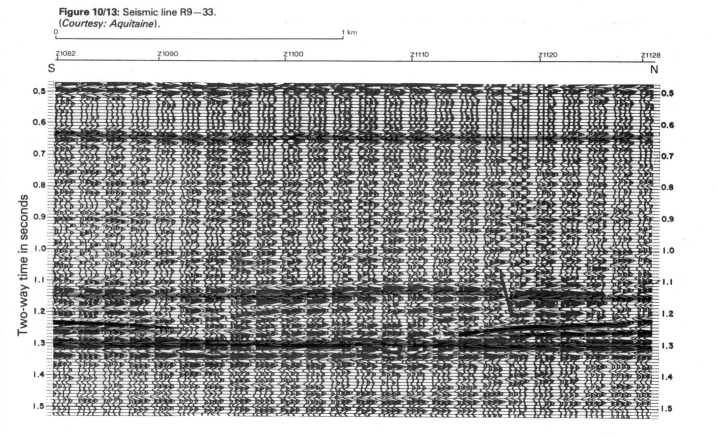

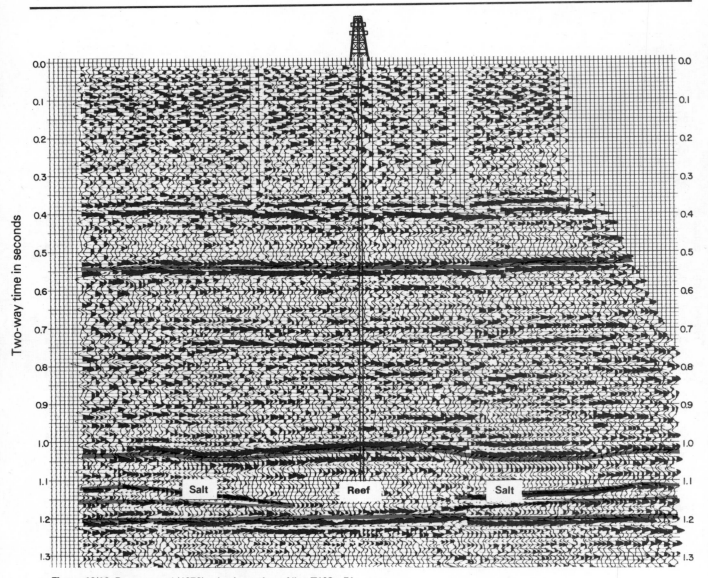

Figure 10/16: Reprocessed (1976) seismic section of line T108—51.

10. RAINBOW LAKE CASE HISTORY

The purpose of this chapter is to describe how, in 1965 and subsequently, the relatively new concept of common depth point reflection seismic technique was effectively used to contribute to the discovery and development of reservoirs in a series of stratigraphic traps which ultimately yielded recoverable reserves of an estimated 2.2 billion barrels of oil and 1.5 trillion cubic feet of gas, and which were collectively known as the Rainbow Lake Fields.*

10.1 Location

The Rainbow Lake area (figure 10/1) is situated in the extreme northwest corner of the Province of Alberta in Canada, 25 miles east of the boundary with British Columbia (120°W longitude) and 100 miles south of the boundary with the Northwest Territories (60°N latitude). It is over 400 miles northwest of the provincial capital, Edmonton, and about 700 miles from Canada's oil capital, Calgary.

10.2 Regional setting and stratigraphy

The Rainbow area is situated towards the northern end of the Western Sedimentary Basin or Interior Plains (figure 10/2). Economic basement is composed of Pre-Cambrian metasediments and igneous rocks.

From the edge of the exposed Pre-Cambrian shield, the basement dips very gently southwestwards for about 500 miles, then rapidly disappears deep beneath the Rocky Mountains where Palaeozoic and Pre-Cambrian rocks are thrust over the accumulated 15,000—20,000ft of Mesozoic and Palaeozoic sediments.

In the Rainbow Lake area, basement is generally around 5,500ft below sea level or at 7,300ft drill depth, and dips are around one quarter of a degree (25ft per mile). The basement is overlain by mainly Devonian evaporites, carbonates and shales. The Middle Devonian is around 2,500ft thick and restricted almost entirely to carbonates and evaporites. The Upper Devonian has a thick, almost 2,000ft section of marls and marly shales overlain by a limestone unit, the Wabamun, 1,200ft thick. A major unconformity exists between the latter and the approximately 700ft thick Lower Mississippian marls and shales and a further unconformity exists below the overlying 1,200ft of Lower Cretaceous shales which are the bedrock of the area. Glacial deposits of extremely variable thickness cover the area and from an operational point of view, surface conditions can be extremely difficult, the surface water and muskeg conditions (mossy growth found in high latitudes, usually as generally uneven topography encouraging standing water and deep and treacherous swamp, bog or marsh). Understandably, both seismic and drilling operations are most efficient during the annual freeze-up (November—April) and indeed in most areas are impossible outside this period.

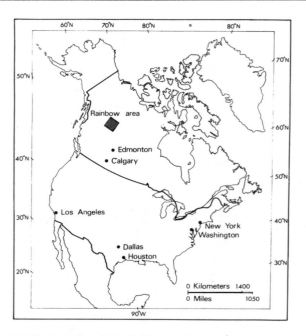

Figure 10/1: Location of the Rainbow Lake area.

The main exploration target for the area is the Rainbow Member of the Keg River Formation where hydrocarbons are trapped in reef structure reservoirs. During the Middle Devonian, the area was a restricted basin in which palaeo-temperatures, winds and currents were favourable for the development of reefs which grew up to 800ft above the main platform of the Lower Member of the Keg River Formation. The basin was restricted by reef barrier complexes to the east, by the Hay River barrier to the southeast, and by the Sheikilie to the west (figure 10/3). The increasing dominance of these barriers resulted eventually in the cessation of reef growth, deposition of salt between the reefs, followed by a continuous anhydrite sequence which serves as the trapping mechanism for the Rainbow Member Reservoirs.

The regional setting is well illustrated by the total intensity magnetic map (figure 10/4). The eastern boundary of the basin is defined by the Hay River Fault which is a major fault system trending NE-SW for hundreds of miles, down-throwing to the northwest, and which is marked by a linear magnetic high of 700 gamma amplitude. In the western part of the map a magnetic low, trending approximately N-S, is probably related to major petrological variations in the basement. As such this zone may have had topographic relief and acted as a hingeline for barrier reef growth.

*Our sincere thanks go to Aquitaine Company of Canada Ltd and Mobil Oil Canada Ltd for permission to use the previously unpublished seismic sections and maps.

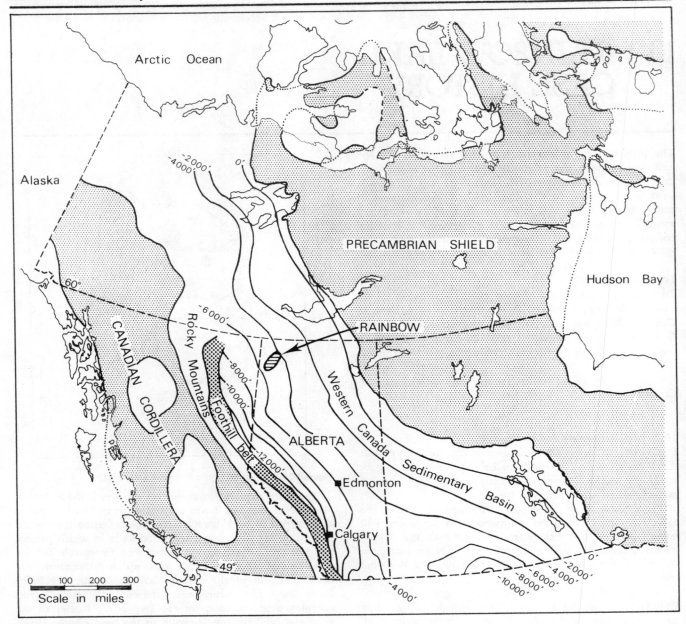

Figure 10/2: Regional geological map.

10.3 Seismic interpretation

In the area there are four main reflectors (figure 10/5), and as a means of defining prospective structures it is necessary to interpret and map all four. These reflectors have been named and can be described geologically as follows:

1. **Wabamun.** This reflector marks the top of the Wabamun unit where there is a major unconformity between Upper Devonian and Lower Mississippian rocks. A large velocity change exists between the shale and marl above the Wabamun unit (8–10,000ft/s) and the underlying limestone (15–16,000ft/s) providing reflection coefficients in the range 0.27 to 0.39. The Wabamun is a relatively homogeneous unit over 1,000ft thick for which a two-way interval time would be 0.125s and thus the reflection is unaffected by succeeding reflections. It is accordingly an excellent reflector in all good data sections.

2. **Slave Point.** This reflector marks the top of the Middle Devonian. Again we have a limestone unit overlain by shale. The shale unit varies in lime content and has velocities in the range 11–15,000ft/s (the velocity contrasts for these and other formations are well illustrated by the sonic logs shown in figure 10/7b). With the Slave Point limestone velocity of approximately 18,000ft/s, reflection coefficients vary from 0.14 to 0.31. On multi-fold seismic sections (figure 10/7a) the reflection is generally good but it is occasionally degraded by either thickness and velocity variations in the overlying shale units or by thinning of the Slave Point limestone (maximum thickness 270ft: equivalent to 0.030s two-way time), in which case there is interference with reflections from underlying formations of varying thickness and lithology. Reflection quality is poor where multiple suppression is inadequate; first-order Wabamun multiples, and at

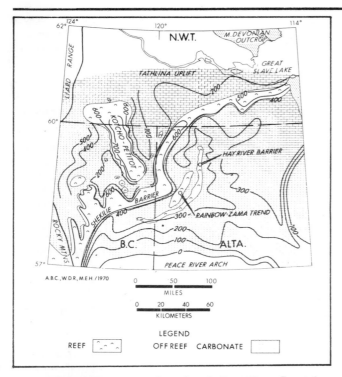

Figure 10/3: Rainbow area, isopach of Keg River Formation showing barrier complexes and reef trend surrounded by off-reef carbonate facies (after Hriskevich, 1970).

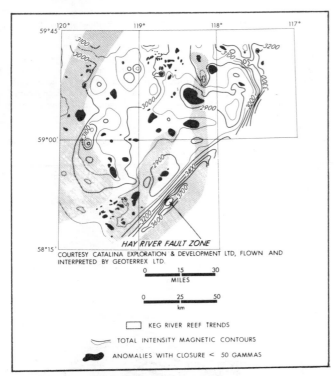

Figure 10/4: Central Rainbow area total force intensity aeromagnetic map (after Hriskevich, 1970).

times the higher-order Cretaceous-Mississippian unconformity multiples, often coincide with the Slave Point reflection. This situation bedevilled earlier single-fold record interpretation (see figure 10/6c) and provided one good reason for the application of multi-fold techniques.

3. **Black Creek Salt.** The salt unit occurs over most of the basin between the reefs, post-dating them, and both top and bottom of this unit provide good reflectors. It lies above the Keg River Formation argillaceous/carbonate units and is overlain by anhydrite and dolomites. Utilising respective velocities of 18,000ft/s, 14,000ft/s and 17,500ft/s and densities of 2.80, 2.25 and 2.60g/cc, we can derive reflection coefficients of −0.23 and +0.18 for the top and bottom reflectors. The Black Creek Salt has a maximum observed thickness of 271ft and thins out completely mainly due to solution around the reefs. However, because of the high reflection coefficients, and with suitable broadband recording and good quality data, it has been possible to pick peak to trough minima as low as 0.007s (49ft) for the top to bottom intervals.

4. **Base Cold Lake Salt.** This salt unit is remarkably consistent throughout the area and is a good reflector. It occurs about 500ft below the base of the Rainbow Member and is generally about 70ft above the Pre-Cambrian basement. Thickness varies from 0−180ft. The base provides velocity contrasts from around 14,000 to 18,000ft/s and one suspects that where maximum amplitudes occur, tuning of the top and bottom is taking place. The event may be up to three cycles in length. In some illustrations it is also described as 'Red Beds'.

In 1964, with the poor quality seismic data then available the selection by Banff Oil* of a prospective location was both speculative and courageous. Only one well had been drilled in the area before; in 1954 the 10−27−109−9 W6M well encountered 245ft of Black Creek Salt and its complete lack of hydrocarbon indications provided no incentive for a follow-up. According to Hriskevich (1970): 'The seismic data on which the discovery well at Rainbow was based were obtained during 1953−1955 before the use of common depth point techniques . . . seismic interpretation was made extremely difficult by the presence of a severe multiple problem. The selection of the location involved very close co-ordination between geologists and geophysicists.' Single-fold seismic records of the type referred to above are illustrated in figure 10/6. Figure 10/6a shows Wabamun, Slave Point and Cold Lake Salt events quite clearly, plus a supposed reef event. Figure 10/6b shows more or less the same section but with a salt event replacing the reef event. Figure 10/6c referred to earlier shows poorer data with severe noise including multiples in the Slave Point reflection window. However, despite these difficulties in seismic interpretation, the well (Banff Aquitaine 7−32−109−8 W6M) was successful in discovering both oil and gas in commercial quantities. Results showed a total net pay 480ft oil and 291ft gas and this acted as a great stimulus to obtain the best possible data in new exploration of the area. Common depth point stacking was by then the order of the day, but other improvements rapidly followed:

* In 1965 the original participants in the exploration play were Banff Oil Ltd (operator), Aquitaine Company of Canada Ltd (a subsidiary of Société Nationale des Pétroles d'Aquitaine, France) and Socony Mobil Oil of Canada Ltd. The working interests were 5, 45 and 50 per cent respectively. Banff Oil was subsequently merged with Aquitaine in 1970.

AGE	FORMATION	THICKNESS	LITHOLOGY
Lower Cretaceous	Lower Cretaceous		
Lower Miss.	Lower Mississippian	700'	
Upper Devonian	Wabamun	1200'	
Upper Devonian	Fort Simpson	2000'	
Middle Devonian	Slave Point	235'	
Middle Devonian	Watt Mountain	45'	
Middle Devonian	Sulphur Point	75'	
Middle Devonian	Muskeg / Black Creek Salt / Rainbow Member	1000'	
Middle Devonian	Lower Keg River		
Lower or Middle Devonian	Chinchaga	275'	
Lower or Middle Devonian	Cold Lake	0'—150'	
Lower or Middle Devonian	Ernestina Lake	40'	
Lower or Middle Devonian	Basal Red Beds	25'	
	Precambrian		

Shale Marl Dolomite Salt
Sandstone Limestone Anhydrite Metamorphics

Figure 10/5: Stratigraphic section.

Survey grid spacing: Closer density grids were surveyed than previously with E—W lines at quarter-mile spacing and with occasional N—S tie lines over identified anomalies.*

Shot patterns: In general, single charges of up to 5lbs were used in holes up to 60ft deep but in glacial drift and other difficult near-surface conditions improvements were achieved by shooting 3-hole patterns.

Spread geometry: Various spread geometries, split-spread or end on, were used recording 24 channels with 110—150ft geophone station spacing. Where necessary, offsets or gaps were utilised to remove ground roll. Shot-hole spacing was 110—150ft for 12-fold coverage, 220—300ft for 6-fold.

Geophone arrays: Generally nine to eighteen phones were used in a 160—170ft string. Geophones were 14Hz natural frequency or less.

---

* In the provinces of Canada, much of the Government surveyed areas are divided into Township areas, 6 x 6 miles square, each Township being sub-divided into 36 sections, numbered from 1 in the SE corner alternatively left and right to 36 in the NE corner. Each square mile section is further sub-divided into 16 l.s.d.s. (legal sub-divisions) numbered in similar fashion from 1 in the SE to 16 in the NE. At the time of the Rainbow discovery, the Alberta Oil and Gas Conservation Board, the Provincial Regulatory Authority, limited oil production wells to within l.s.d.s. 2, 4, 10 and 12, providing 160 acre per well drainage units, except in special cases. Accordingly it was good practice to shoot seismic lines through the centre of those l.s.d.s. on a one-half mile spacing. With this type of programme layout, new wells could be staked at a specific shot point location, as near to the reef crest/maximum pay point. The coverage was subsequently doubled to one-quarter mile spacing to ensure proper delineation of reef anomalies that were sometimes less than three-quarters of a mile across. Thus it can be seen that planning of seismic survey grids in this area was to a large extent influenced by legal as well as geological and operational considerations.

Figure 10/6: Rainbow area seismic field records, 1953 (after Hriskevich, 1970).

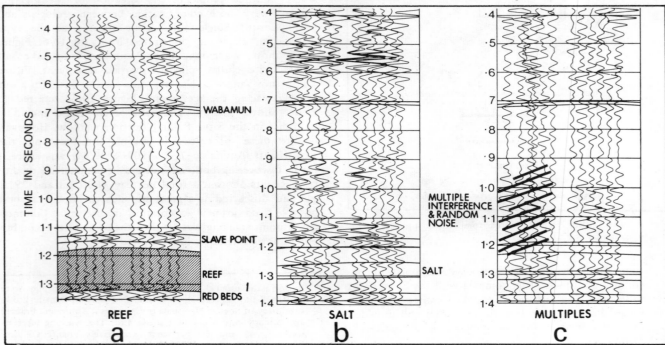

a — REEF — WABAMUN, SLAVE POINT, REEF, RED BEDS

b — SALT — WABAMUN, SALT

c — MULTIPLES — MULTIPLE INTERFERENCE & RANDOM NOISE.

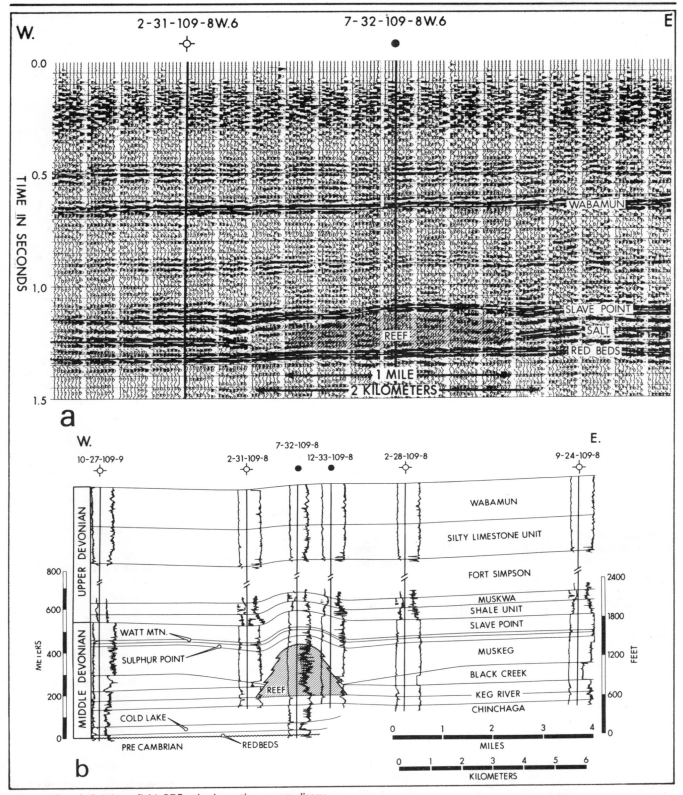

Figure 10/7: a) Rainbow field CDP seismic section across discovery pool (after Hriskevich, 1970). **b)** Rainbow field geological cross-section across discovery pool (after Hriskevich, 1970).

Recording systems: The seismic 'boom' at Rainbow coincided with the introduction of the first digital field systems, and many crews were equipped with this. Record filter passbands were generally in the range 14 to 65 or 72Hz. Use of digital systems permitted the application of more advanced processing techniques leading to a significant enhancement of data quality.

Processing: Analogue processing included the application of weathering corrections, elevation statics and, optionally, manual statics to artificially 'flatten' the Wabamun or preferably, the Base Cold Lake Salt reflections. (The Wabamun refracted at about a half-mile offset and consequently only half the traces could be used). The flattening enhanced the quality of the stacking and aided the interpreter in measuring isochrons on the sections. In addition to the same weathering and elevation statics corrections the use of digital techniques allowed computer centres to provide automatic static-picking routines which could either 'structure-smooth' or 'flatten' a key reflection.

The result of applying these techniques was that data quality was greatly improved and the first-order multiple problem, referred to earlier, was easily resolved.

As exploration drilling accelerated in pace with the expansion of seismic programmes, it became apparent that the reef reflection, as shown on the single-fold record of figure 10/6a was not consistent enough to be relied upon as a diagnostic of reef occurrence and could occasionally be confused with the salt reflections. Careful analysis of seismic data did however allow development of an indirect method of reef identification which was based on the following criteria:

Reef presence criteria: Rainbow area

1. Thickening of the Slave Point to Base Cold Lake Salt isochron of the order of 0.015 + seconds.
   and/or
2. Thinning of the Wabamun to Slave Point isochron of the order of 0.015 + seconds.
   and
3. Absence of the easily identified Black Creek Salt event between the Slave Point and Base Cold Lake Salt events.

Figure 10/7a illustrates the principles well. It is an E—W section through the discovery well, Banff Aquitaine Rainbow West 7—32—109—8 W6M. In this section the Red Beds/Base Cold Lake Salt reflection has been artificially 'flattened' (datumised) for maximising stacking efficiency. This and other tying sections on this anomaly comply with the criteria listed above; in particular it should be noted that the prominent Black Creek Salt Reflection disappears from the west one trace-panel east on the 2—31 well and can be seen again 5 trace-panels from the eastern end of the section (criterion 3). Confirmation of the salt presence to the west is provided by the 2—31—109—8 well which penetrated 80ft (0.011s) of salt. The geological cross section in figure 10/7b summarises the sub-surface geology. The structural anomaly at the Slave Point level overlies what is known as the 'A' pool; it covers an area of only 985 acres or 1.5 aquare miles yet the reef is 756ft high and has 686 gross feet of oil and gas pay and a reservoir of 164 MM bbls.

Reefs at Rainbow come in all shapes and sizes. Figure 10/8a is an E—W line over two reefs. The 'D' pool on the left of the section is only 400 acres or 5/8 square miles. The 'B' pool anomaly is large enough, 3,640 acres or approximately 5 square miles, to be termed an atoll reef with outer rim and inner lagoon. All the interpretation criteria are met for both reef structures. However, there are some anomalous isochron values for the 'B' pool related to the rim and lagoonal developments. Basement faults

with throws of up to 80ft are noted in the area, and this would account for the thinning of around 0.009—0.010s on the upthrown side of the fault seen in figure 10/8b.

An interesting additional criterion which may be used to aid reef identification is by frequency analysis as described by Lindseth.* In figure 10/9 (see p. 142) a frequency analysis of a fixed time-window is provided at the seismic section (basically the same line as figure 10/7a). A marked frequency shift from 37 to 47Hz can be noted from left to right over the Keg River Reef Structure. The reason for this frequency shift would appear to be that compaction of the sediments draped over the reef structure produces thinning of beds, and consequently high frequency 'tuning' effects.

10.5 Interpretation and mapping of a Rainbow Reef

Reconnaissance seismic line T108—51 shown in figure 10/10 located a reef anomaly and additional lines T108—44 (figure 10/11) R9—32 (figure 10/12) and R9—33 (figure 10/13) confirmed its areal extent (see pp. 144-145).

The Wabamun, Slave Point (2nd leg), Top and Bottom of the seismic events for Black Creek Salt, and the Base Cold Lake Salt are identified. T108—51 (figure 10/10) is, by processing, 'flattened' on the Base Cold Lake Salt event. It shows typical 'dead' reef character, 'roll-over/drape' on the Slave Point of around 0.030s, and Black Creek Salt character disappearing at shot-point V16567 and reappearing at V16593. The east side of the reef appears to have steeper relief. T108—44 (figure 10/11) is flattened on the Wabamun. Roll-over on the Slave Point is 0.023 to 0.035s west to east and the smooth Base Cold Salt reflector shows a slight high below the reef. It is apparent that the reef has a steep western margin and that the Black Creek Salt character disappears at shot-point V17362, reappearing at V17395.5. R9—32 (figure 10/12) is flattened on the Base Cold Lake Salt. Roll-over is over 0.030s and the steep relief to the north is accentuated by an apparent fault in Slave Point at shot-point Z1168 which may be induced by both Salt removal and the flattening of the same fault at Base Cold Lake Salt level. The Black Creek Salt character disappears at that point and reappears at shot-point Z1298. R9—33 (figure 10/13) is also flattened on the Base Cold Lake Salt. It crosses the west flank of the reef. Consequently, the roll-over is in the region of 0.02s: however, the steep relief to the north is obvious and a fault in the Slave Point is apparent at shot-point Z1118. (Note that the actual mapping 'picks' are made one trace left of the actual shot-point location. For convenience in mapping, the faults have been dislocated one shot-point to the north). From a thickness of 0.021s or 144ft, at shot-point Z1128, the Black Creek Salt disappears at 1117 and reappears at 1091.

Mapping of the two main, complementary isochrons is shown in figures 10/14 and 10/15. There is a minimum closure of 0.023s on the Wabamun to Slave Point isochron and 0.029s on the Slave Point to Base Cold Lake Salt isochron. As it is known that there is negligible structural relief on either the Wabamun or Base Cold Lake Salt, it can be assumed that the respective thinning and thickening reflects structure on the Slave Point, with perhaps some compaction-induced interval velocity increase effect in the

---

* 'Recent Advances in Digital Processing', 1968, unpublished report.

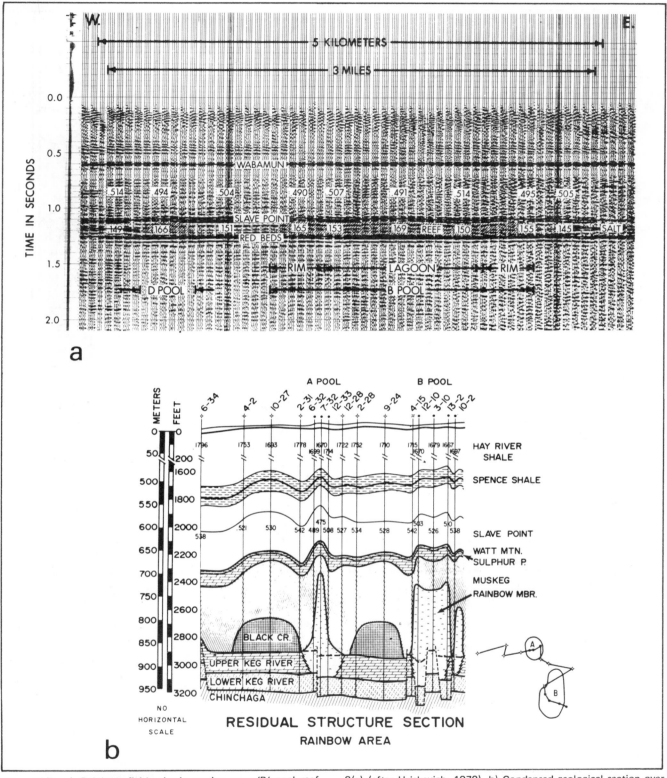

a

b

RESIDUAL STRUCTURE SECTION
RAINBOW AREA

Figure 10/8: a) Rainbow field seismic section over 'B' pool reef showing outer rim, inner lagoon and 'D' pool pinnacle reef. Note anomalously thin Slave Point to Base Cold Lake Salt (Red Beds). interval on eastern rim related to faulting of type shown in figure

8(a) (after Hriskevich, 1970). b) Condensed geological section over 'A' and 'B' pools. Note pronounced faulting in the lowermost section and compare with figure 8(a) (after Barss *et al*, 1970).

first isochron: evidence from wells shows that the anomaly is almost entirely a drape effect. The third reef criterion, absence of Black Creek Salt, is indicated by the zero edge of the Salt character. The four lines tie perfectly and their

isochron values are posted. For clarity, additional lines used for the surrounding contour control are shown without their values. This reef is an outstanding anomaly: using the 0.150s Slave Point to Base Cold Lake Salt

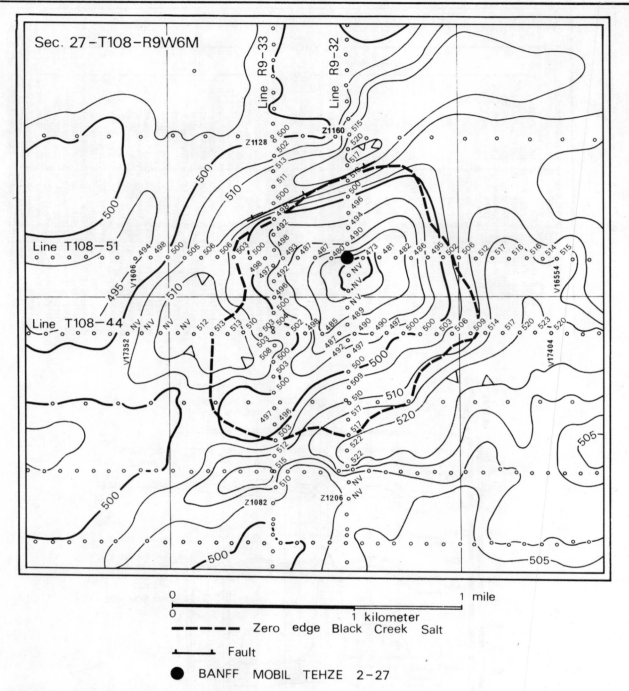

Figure 10/14: Map of two-way reflection time interval between Wabamun and Slave Point horizons. (*Courtesy: Aquitaine*).

contour, the maximum diameter is under a mile. The discovery well Banff Mobil Tehze 2–27–108–9 had 958ft of reef, 595ft of hydrocarbon pay and reserves of 27 MM bbls oil and 13 BCF natural gas were proven in the reef and overlying carbonate units.

Figure 10/16 shown on p. 146 is a 1976 reprocessing of Line T108–51. Significant improvement of reflection quality is achieved through more refined normal moveout and static corrections plus deconvolution; the anomaly is further heightened by a condensed horizontal scale (12 traces per inch versus 8 per inch and a 1 trace gap every two shot-points on the original presentation).

The spate of discoveries and further government lease sales in the Rainbow Lake region at one time reached a level such that development plans required geophysicists to double as explorationists and productionists. In developing well prognoses for reservoir delineation and production, it soon became evident, on a localised basis, that there were simple empirical relationships between the Slave Point drape in time and depth, and the amount of vertical relief on the Rainbow Member. Careful analysis of these relationships eventually resulted in prognosis accuracies in the order of 50ft or less, or within 1 per cent of target drill-depth.

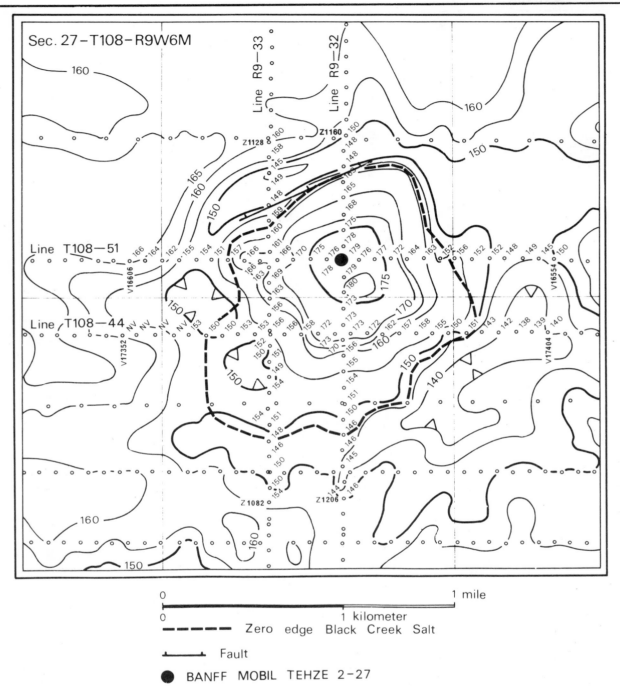

Sec. 27 – T108 – R9W6M

Line R9—33
Line R9—32
Line T108—51
Line T108—44

0   1 mile
0   1 kilometer
– – – – – Zero edge Black Creek Salt
⊢—— Fault
● BANFF MOBIL TEHZE 2–27

Figure 10/15: Map of two-way reflection time interval between Slave Point and Base Cold Lake Salt horizons. *(Courtesy: Aquitaine).*

With methodical seismic mapping and tying in of well data, it was eventually possible to prepare maps on the Keg River Reef/Rainbow Member itself. A simplified three-dimensional view of these reefs is shown in figure 10/17, and in figure 10/18 a schematic section is shown including data on the reservoir characteristics. As mentioned earlier it was seismically possible to map the top and bottom of the Black Creek Salt, and given the homogeneous half-velocity of 7000ft/s, derive the isopach. Using 100ft contours, a map of the complete Rainbow Black Creek Salt Basin is shown in figure 10/19. These last two figures are part of the end product of analysis of a vast amount of data, interpreted by many individuals and a number of companies. The history of exploration and development of the Rainbow area has depended on two main factors. First, the initial successful interpretation of old data which led to the discovery of oil and gas by Banff Aquitaine and second, the fact that this coincided with a great technical advance in seismic exploration methods, common depth point stacking and digital recording. This coincidence led to an immediate and major expansion in Alberta exploration effort, both seismic and drilling.

155

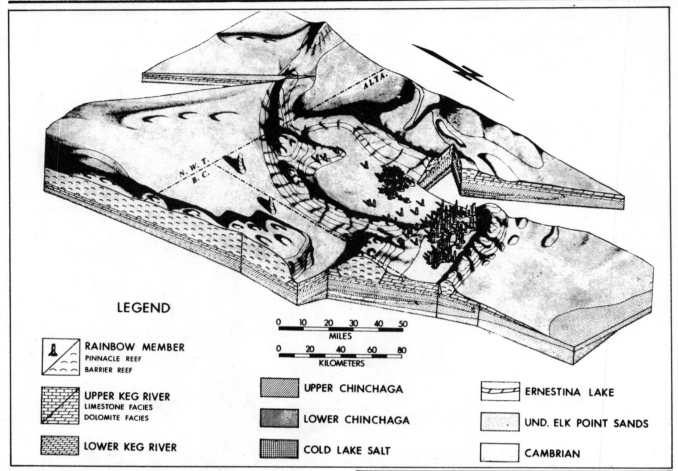

LEGEND

RAINBOW MEMBER
PINNACLE REEF
BARRIER REEF

UPPER KEG RIVER
LIMESTONE FACIES
DOLOMITE FACIES

LOWER KEG RIVER

0   10   20   30   40   50
MILES

0   20   40   60   80
KILOMETERS

UPPER CHINCHAGA

LOWER CHINCHAGA

COLD LAKE SALT

ERNESTINA LAKE

UND. ELK POINT SANDS

CAMBRIAN

Figure 10/17: 3-D representation of Keg River Reef formation (after Barss *et al*).

The end result, coming in only five years, was that all the reef reservoirs shown in figure 10/17, were discovered with a 50 per cent drilling success ratio, providing the establishment of reserves of 2.2 billion barrels of oil and 1.5 trillion cubic feet of natural gas in the Rainbow and associated Zama North area.

In closing it is interesting to note the similarity of the interpretation criteria used in a more recent successful reef play in Northern Michigan in the USA.* Targets are in the Silurian Niagaran reefs. As shown on the stratigraphic columns of figure 10/20, the off-reef section is character-ised by the presence of salt. Also, as at Rainbow, a promin-ent shale/limestone reflection ('Dundee') is used as a datum marker to demonstrate isochron thinning/drape over the reef. The similarities of the interpretation methods are modified due to the thinner salt and carbonate beds in the Niagaran play and are summarised by the comparison table below.

Seismic sections from Rainbow and Northern Michigan are shown in figure 10/21. Despite the difference in data quality, the similarity is strong: note especially the abrupt change of character going from off-reef to on-reef in both cases.

*W.G. Caughlin, F.J. Lucia and N.L. McIver 'The detection and development of Silurian reefs in Northern Michigan' *Geophysics* 41 (1976) No. 4 pp. 646–658.

Reef presence criteria — comparison between Rainbow & Michigan

	Rainbow Area	*Michigan Area*
Isochronal Thinning:	Wabamun to Slave Point	Dundee to A2 carbonate
Isochronal Thickening:	Slave Point to Base Cold Lake Salt	A2 carbonate to Niagaran (generally the Niagaran itself is not seen because of the lack of reflectivity: this itself is an alternative criteria).
Reef Zone Character Change:	Disappearance of Black Creek Salt	Weakening or disappearance of A2 carbonate (the thin salt and carbonate beds encourage a strong 'tuned' event when present: the absence of the salt induces 'detuning').

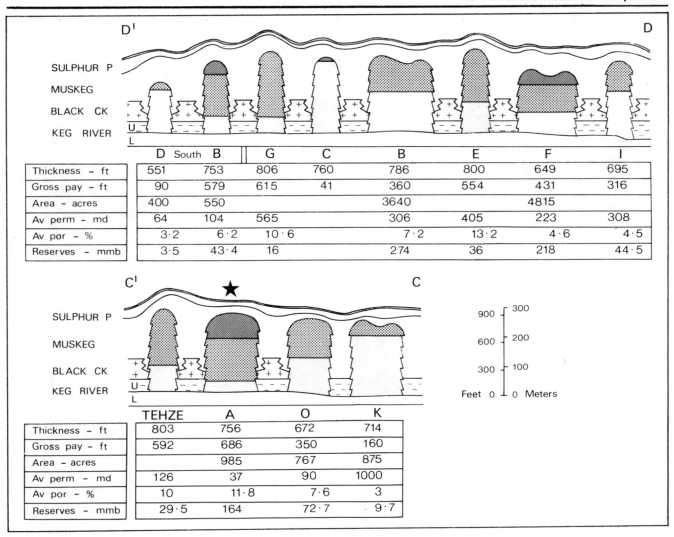

	D South B		G	C	B	E	F	I
Thickness – ft	551	753	806	760	786	800	649	695
Gross pay – ft	90	579	615	41	360	554	431	316
Area – acres	400	550			3640		4815	
Av perm – md	64	104	565		306	405	223	308
Av por – %	3·2	6·2	10·6		7·2	13·2	4·6	4·5
Reserves – mmb	3·5	43·4	16		274	36	218	44·5

	TEHZE	A	O	K
Thickness – ft	803	756	672	714
Gross pay – ft	592	686	350	160
Area – acres		985	767	875
Av perm – md	126	37	90	1000
Av por – %	10	11·8	7·6	3
Reserves – mmb	29·5	164	72·7	9·7

Figure 10/18: Schematic section through Rainbow area reefs showing comparative data on reservoir characteristics.

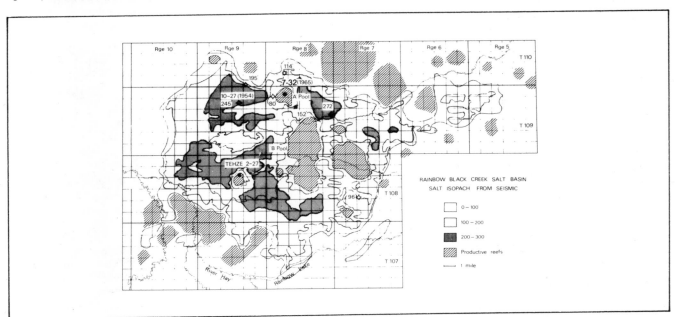

Figure 10/19: Salt isopach map of the Rainbow Black Creek Salt Basin.
(*Courtesy: Mobil 1967*).

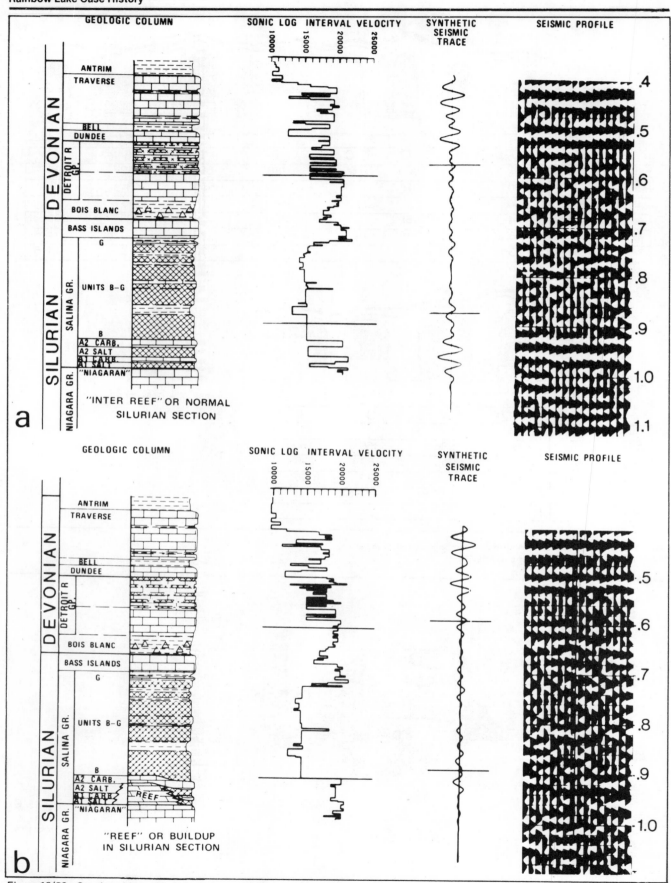

Figure 10/20: Stratigraphic sections in North Michigan compared with sonic logs, synthetic seismic traces and seismic profiles. a) Inter-reef situation; note thin bed tuning due to alternations of salt and carbonates in lower Salina zone. b) Reef situation: note lack of reflectivity in lower Salina zone due to loss of salt beds (after Caughlin *et al*, 1976).

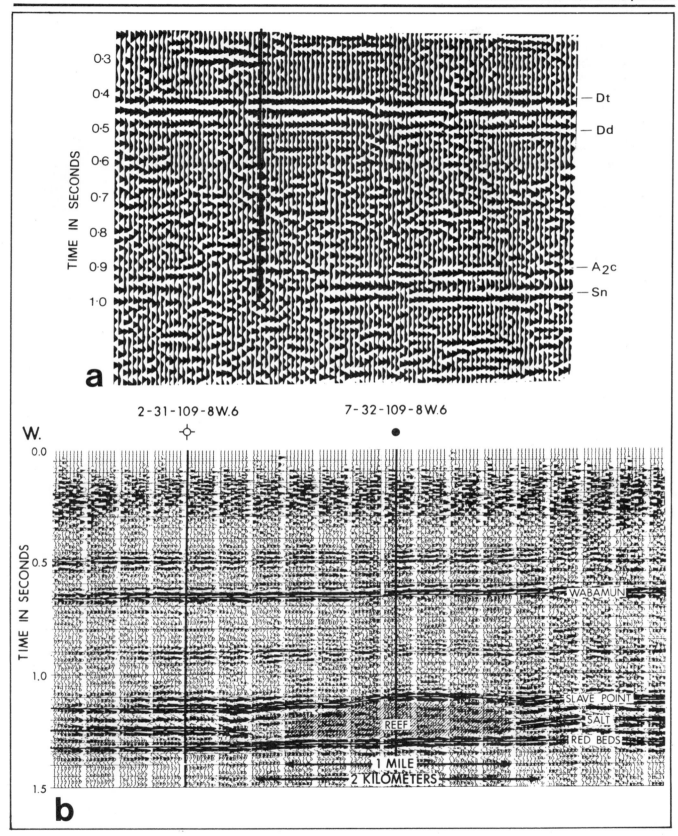

Figure 10/21: Comparison of seismic sections across reefs in North Michigan and Rainbow. **a)** Reef in North Michigan: note loss of reflectivity in reef zone beneath Well 1–31. **b)** Section across Rainbow field discovery pool: note similar loss of reflectivity beneath Well 7–32–109–8W.6. (Section a) after Caughlin *et al* 1976; section **b)** after Hriskevich, 1970).

References

D.L. Barss, A.B. Copland and W.D. Ritchie, 'Geology of the
      Middle Devonian reefs, Rainbow area, Alberta,
      Canada', in Halbouty, M.T. *(ed). Geology of giant
      petroleum fields.* Mem. AAPG, No. 14, pp.19-49
      (1970).

W.G. Caughlin, F.J. Lucia and N.L. McIver, 'The detection
     and development of Silurian reefs in Northern
     Michigan'. *Geophysics,* 41 (1976) No.4, pp 646-658.

M.E. Hriskevich, 'Middle Devonian reef production,
      Rainbow area, Alberta, Canada', *Bull. AAPG,* 54
      (1970) No. 12, pp. 2260-2281.

11. GEOPHYSICAL CASE STUDY OF THE KINGFISH OILFIELD OF SOUTH-EASTERN AUSTRALIA

The Kingfish Oilfield is located in the Gippsland Basin of southeastern Australia. It lies approximately 50 miles offshore in 250 feet of water and is currently Australia's largest producing oilfield with an estimated ultimate recovery of approximately one billion barrels (figure 11/1).

The oil is trapped within a large east-west trending Late Eocene palaeotopographic high. The top of the reservoir is an unconformity surface which is sealed by fine grained marine clastics of Oligocene age. The reservoir itself consists of Lower Eocene and Upper Palaeocene deltaic and marginal marine sandstones of the Latrobe Group.

Large lateral variations in the average velocity to the top of the reservoir have complicated the seismic mapping of this field. This chapter will discuss the origin of the velocity

problems, the magnitude of their effects, and the practical techniques which have been used to obtain velocities for reliable time-depth conversions.

11.1 Regional stratigraphy of the Gippsland Basin

The regional stratigraphy of the Gippsland Basin is summarised in a schematic cross-section which runs from north to south across the basin (figure 11/2). All the commercial hydrocarbons discovered in the basin occur in sands within the Upper Cretaceous to Eocene Latrobe Group which is a thick sequence of predominantly non-marine fluvial and deltaic coarse clastics. The overlying Lakes Entrance formation is comprised of calcareous mudstones and shales which were deposited during a widespread marine transgression of the basin during the

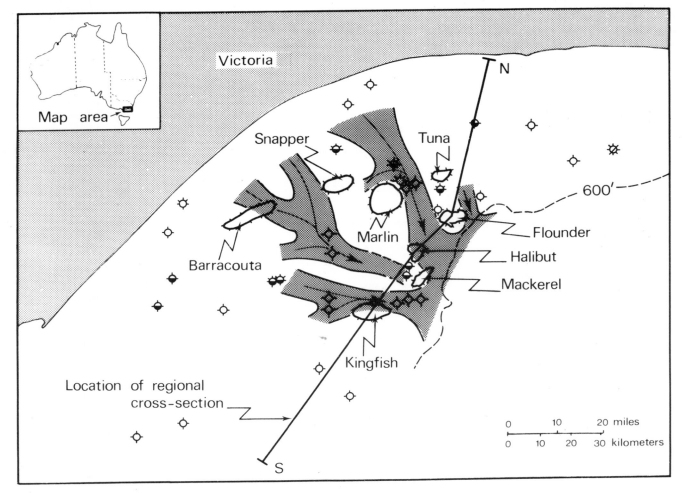

Figure 11/1: Gippsland Basin location map showing the major oil and gas fields and the Miocene channel systems.

Oligocene. Marine sedimentation continued through the Miocene to Recent with the deposition of skeletal limestones, calcarenites and marls which are referred to as the Gippsland Formation.

Sea level changes and structural movements during Miocene time triggered the development of a complex system of large scale submarine canyons. The sedimentary fill within these channels gives rise to the principal velocity problems experienced in the basin. At the head of each channel the fill is a mixture of coarse skeletal fragments, but in the middle and distal portions of the channels the infilling sediment is a dense micritic limestone. The interval velocity of this micritic limestone may be as high as 15,000ft/s, compared to the 8,000 to 9,000ft/s interval velocity of the surrounding shales. This large velocity contrast results in significant lateral variations in the average velocity to the underlying seismic markers and these variations severely distort the seismic reflection times to any underlying structure.

Figure 11/1 shows the distribution of the major Miocene channels. They originate in the northwestern part of the basin and trend to the southeast where they coalesce to the east of the Kingfish-Mackerel area. It should be noted that these channels pass across most of the major oil and gas fields which are therefore affected to some degree by the resulting velocity problems.

11.2 Historical resumé

The first seismic survey in the offshore portion of the Gippsland Basin was conducted in 1962 and comprised single fold, split spread, analogue reconnaissance data. The Kingfish structure was recognised on two lines from this survey.

Prior to the drilling of Kingfish-1, two subsequent seismic surveys provided eight more lines of seismic control over the structure. These lines consisted of six-fold, split spread, analogue data and revealed a large channel complex in the Miocene section above the Kingfish structure. A similar Miocene channel system was known to cause lateral velocity variations in the vicinity of the Marlin field, 25 miles to the north. It was considered, however, that any velocity variations which might be present in the Kingfish area would not be sufficient to cause drastic modification of a structure which was mapped as having 100 msec of closure over 50 square miles. In addition, the accuracy of velocities derived from $T \overset{\triangle}{} T$ analysis of split spread data from a relatively short cable was suspect and the use of such velocities in depth conversion was considered just as likely to produce depth errors as to correct them.

The first well, Kingfish-1, was therefore drilled on the time crest (figure 11/3). The well intersected 114ft of oil trapped immediately beneath the top of the Latrobe Group and the velocity information from the well indicated an oil-water contact at a reflection time of 1.590s. Relating this oil-water percentage trap fill and an estimated 200 million barrels of oil-in-place. This assessment was disappointing considering the size of the closure evident from the seismic reflection data.

However, the interval velocities derived from the sonic log showed a more severe velocity problem than had been anticipated (figure 11/4). The channel sediments had an interval velocity of 13,000ft/s which was significantly faster than anticipated in the pre-drill analysis. The interval velocity of the underlying and surrounding Lakes Entrance shales was 9,000ft/s or 4,000ft/s slower than the channel fill. It was realised that the reduction in travel time in those areas where there was a thick high velocity channel fill would cause a considerable distortion of the underlying time structure at the top of the Latrobe Group. It was also apparent that a more comprehensive seismic programme with the emphasis on obtaining velocity data was warranted in order to more accurately delineate the Kingfish structure prior to any additional drilling. Consequently, ten digitally recorded six-fold CDP seismic lines

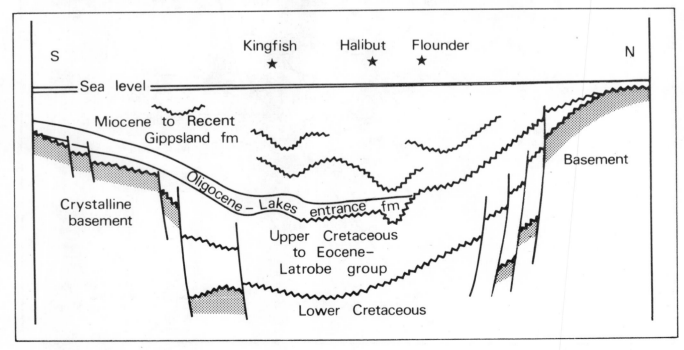

Figure 11/2: Diagramatic cross section of the Gippsland Basin along the line S—N figure 11/1.

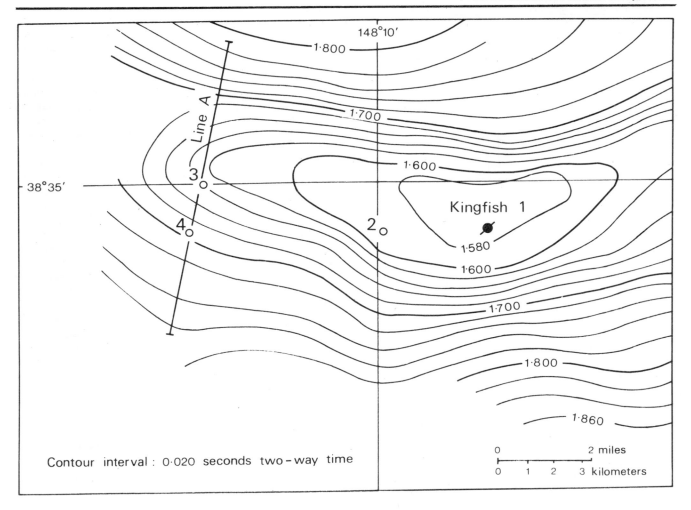

Figure 11/3: Two-way reflection time map of the top of the Latrobe Group at Kingfish.

were shot across the structure. As suspected, intensive $T \Delta T$ velocity analysis of this data confirmed the existence of a severe velocity gradient in the area and this gradient conformed closely to the form and trend of the base of the Miocene channel system.

From the velocity data, it became apparent that the Kingfish-1 well had been drilled on the flank of the field only one half mile from the eastern limit of the oil-water contact. In fact, due to the severity of the velocity gradient across the area, the revised crest of the structure was interpreted, at this time, to be two miles west of the time crest.

To verify the revised structural interpretation of the reservoir, Kingfish-2 was drilled in late 1967 at a location 11,000ft to the west of the Kingfish-1 well. This stepout well, which intersected the top of Latrobe Group 118ft shallower than Kingfish-1, penetrated 232ft of gross oil sand above the same oil-water contact as seen in the discovery well.

A further stepout well was considered necessary to delineate both the structural configuration and stratigraphy of the western portion of the reservoir and in early 1968 Kingfish-3 was drilled at a location some six miles to the west of Kingfish-1.

Following the drilling of Kingfish-3 as a successful confirmation well, it was apparent that the three wells tied to the available seismic grid were adequate to provide reliable definition of an oil reservoir having reserves of approximately one billion barrels. In May 1968 the Kingfish oilfield was declared commercial.

The development of the Kingfish field was based on two 21 conductor platforms. The 'A' platform was located approximately 7,000ft west of Kingfish-2 and the 'B' platform approximately 5,500ft east of Kingfish-2 (see figure 11/7). Development drilling began in March 1970 and continued until October 1971. Both platforms were brought on stream during 1972.

Subsequent to the development, three additional stepout wells have been drilled. Kingfish-4 was drilled near Kingfish-3 in 1973 to clarify the stratigraphy of the western end of the field. This well also confirmed the structural picture of this area. In late 1974 and early 1975, Kingfish-5 and Kingfish-6 were drilled to the east of Kingfish-1 in an attempt to define the eastern extremity of the oil. Both of these wells failed to encounter hydrocarbons and discredited the detailed structural interpretation of this area of the field. The implication of these East Kingfish results will be discussed later.

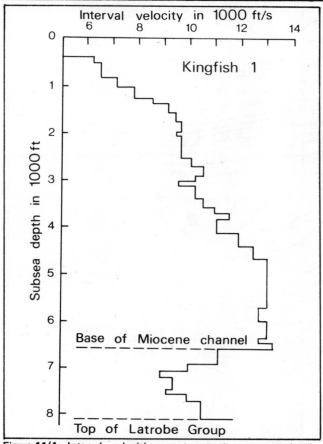

Figure 11/4: Interval velocities measured in the post-Latrobe section of the Kingfish-1 well.

11.3 General geometry of the Kingfish Field

As discussed above, the relationship between the time structure at the top of the Latrobe Group and the apparent structure seen on the seismic sections is complicated by variations in the average velocity caused by the high interval velocity of the Miocene channel fill. This effect is illustrated on the seismic line A (figure 11/5) which crosses the western part of the Kingfish field. On this section, the top of Latrobe time crest of the field can be seen immediately below the axis of the Miocene channel and about one mile north of the Kingfish-3 location. The well control on this line places the true crest two miles south of the time crest, near the Kingfish-4 well. The time crest on this seismic section is in fact about 180ft downdip from the true structural crest.

Figure 11/6 is a time structure map on the base of the Miocene channel system over Kingfish and shows an east-west axis passing north of the Kingfish-2 and Kingfish-3 wells. This map shows that the relief of the channel in the Kingfish area is about 0.200s two-way time, or approximately 1,200ft.

The relationship between the axis of the Latrobe time structure and the axis of the Miocene channelling is obvious. The time structure of the top of the Latrobe Group has been severely distorted by the velocity effects of the channel. Correction for the velocity gradient induced by the channel results in the structural interpretation shown in figure 11/7. This map shows that Kingfish-1, in drilling the time crest, was actually 160ft downdip and five miles east of the true crest of the structure. In fact, the Kingfish-1 well came rather close to completely missing a billion barrel oilfield!

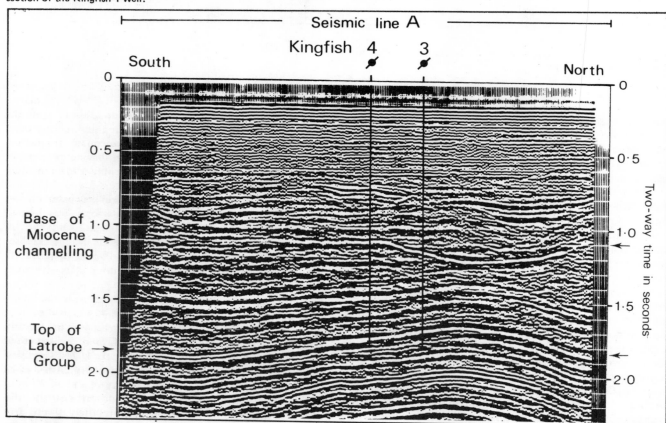

Figure 11/5: Seismic line A (see figure 11/3) across the Kingfish field.

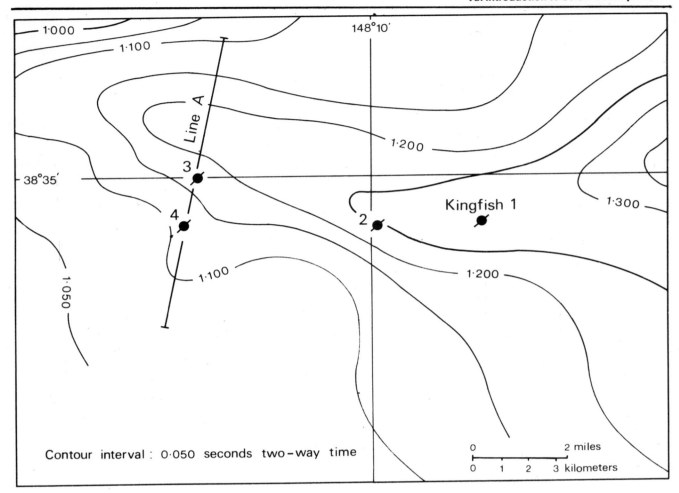

Figure 11/6: Seismic reflection two-way time to the base of the Miocene channel system at Kingfish.

In figure 11/8, the oil-water contact from the Latrobe structure map has been superimposed on the time map and this shows the relationship between the time picture and true structure more clearly. Of particular interest is the extent of the oil-water contact in the southwestern part of the field. If seismic reflection times had been converted to depth using only the Kingfish-1 time-depth information, then the extreme southwestern sector of the field would have been interpreted as almost 1,000ft below the contact.

The Miocene channel system, with its variable high velocity fill, extends over a large part of the Gippsland Basin, as shown in figure 11/1, and many other areas have the same general velocity problem as Kingfish. Exploration experience in the basin has shown that some apparent structures seen on seismic reflection data are entirely velocity induced, whereas others, like Kingfish, are highly distorted by anomalous velocity gradients. Some real structures are not apparent at all on the reflection data. Consequently, every prospect evaluation involves the accurate mapping of average velocities to the top of the Latrobe Group in order to obtain a useable structural picture. As there is not enough well control to do this mapping directly, it is necessary to rely upon indirect velocity determinations from common depth point seismic data to provide the bulk of the information.

11.4 Velocity analysis

11.4.1 Normal moveout velocities and scattergrams

In most offshore areas it is now a relatively simple matter to obtain good quality normal moveout velocities from the routine processing of common depth point seismic data. The general technique used by the processing contractors is to apply an assumed normal moveout to a common depth point gather and then to measure the coherency (a measure of correlation between traces). The normal moveout velocities derived in this manner would only correspond to true average velocities in the unlikely circumstance that the earth was homogeneous and had a constant velocity down to the particular reflection of interest, in which case the seismic rays would have travelled along straight ray paths. Although the moveout velocities are precisely what is required to correct for the effects of moveout during processing, these functions are not of particular use, in the raw state, for converting seismic reflection times to geological depths. In order to obtain useable velocities for this purpose, it is necessary to adjust the moveout velocities for the effects of refraction of the seismic rays as they pass through the layered earth. The remainder of this chapter will be directed towards this problem.

Figure 11/9 shows the type of seismic velocity data that is currently being used in the Gippsland Basin. This 'scattergram' is an output from the '700 package' velocity program provided by Geophysical Service International

(GSI). In the Gippsland Basin such scattergrams are routinely generated every 600 metres on every seismic line giving a high density of velocity control. The velocity program scans for velocities at alternate depth points (every 50 metres) along each seismic line and the scattergram output is the averaged velocities from groups of 12 such depth points. A search for continuity in time, amplitude, dip, and ΔT of events helps to ensure that the velocities that are averaged come from individual continuous reflections. This statistical approach reduces the effect of the random scatter which is prevalent in single depth point velocity determinations.

The scattergram itself is basically a statistical plot of normal moveout velocity against seismic reflection time. The different symbols on the plot represent a ranking of the data points. It should be noted that the prime ranking hierarchy of these points is not based on correlation amplitude but rather on the number of depth points over which the particular seismic event or 'segment' can be followed by the computer. The circles indicate the highest amplitude correlations within each 100 msec gate.

The seismic display on the left hand side of the scattergram is a gather from a central depth point. This gather has been corrected for normal moveout by the application of a previously chosen reference velocity function which is shown plotted on the scattergram. The reference display is a useful aid in the interpretation of the scattergram.

The interpretation of the scattergram involves the drawing of a time-velocity curve which is the best fit to those data points which are considered to be reliable. Obviously, data from multiples and reverberations, etc may appear on the scattergram, and these need to be recognised and discounted in the interpretation. The normal moveout velocity to any particular reflection can be read from the interpreted scattergram and these raw velocities can then be used to determine the average velocities needed to convert seismic reflection times to depths. Specifically, in the Gippsland Basin, it is necessary to determine the average velocity to the top of the Latrobe Group which is the principal objective horizon.

11.4.2 Conversion factor approach to depth conversion

Prior to 1975, the paucity of reliable velocity picks above the base of the Miocene channelling precluded the use of the Dix Approximation to derive average velocities to the top of the Latrobe Group. Instead, a method for depth conversion was adopted which merely involved the application of a percentage correction to the smoothed moveout velocities to produce 'true' average velocities. The smoothing or editing of the moveout data and the selection of appropriate 'conversion factors' are discussed below.

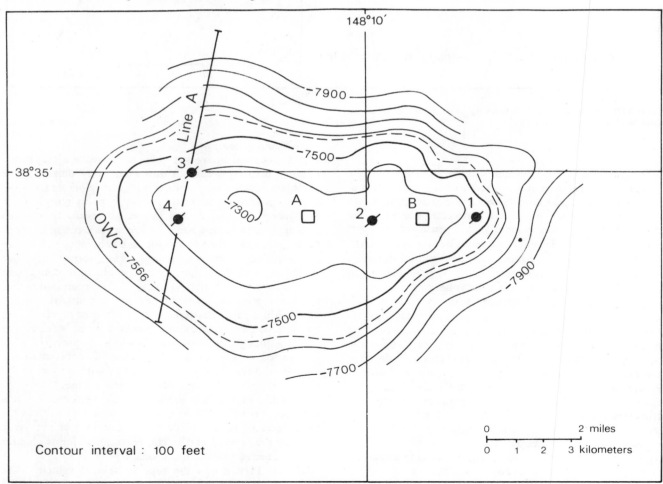

Figure 11/7: Structure map on top of the Latrobe Group at Kingfish. Depth contours in feet.

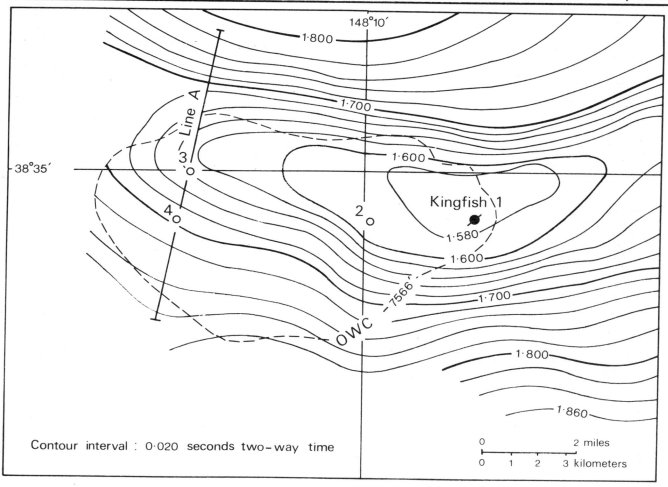

Figure 11/8: Relationship between the top Latrobe time map and the oil-water contact of the Kingfish field.

In situations where the sedimentary section consists of horizontal layers, it is found that reflections from common depth point gathers are approximately hyperbolae on time versus offset plots. In these cases, moveout velocities have a relatively simple relationship to the true average velocities. However, the channelled areas of the Gippsland Basin represent a situation which departs significantly from the horizontally layered case and the channels can produce large variations in normal moveout velocity which are not related to average velocity changes. The origin of these 'geometric effects' are illustrated in figure 11/10a. This illustration models a flat Latrobe surface at 5,000ft with an overlying Miocene channel filled with sediments having an interval velocity of 12,000ft/s. Seismic ray paths have been traced through this model to generate six common depth point 'gathers' and two of these are shown. It can be seen that each of the ray paths contributing to a common depth point set will traverse a different thickness of high velocity material. The moveout curve is significantly non-hyperbolic with the outer rays arriving either too soon or too late depending on the position of the gather relative to the channel geometry. When a distorted moveout curve is forced by a velocity analysis programme to fit a hyperbola, the implied moveout velocity for the central shotpoint of the gather will be either faster or slower than the corresponding moveout velocity for a 'layer-cake' model.

It should be noted that such distorted moveout velocities are quite suitable for moveout correction prior to the stacking of common depth point data, but if the distorted velocities were to be used as the basis for depth conversion, an erroneous picture would result.

Figure 11/10b illustrates the normal moveout velocity profile for the top of the Latrobe Group that results from this channel distortion. The dashed rms (zero-offset) velocity profile has also been plotted to demonstrate how the average velocity to the Latrobe surface should vary. The diagram shows that the moveout velocity is considerably faster than the rms velocity in the vicinity of the channel edges and is slower beneath the channel axis.

Figure 11/10c shows the reflection time profile of the Latrobe surface which is flat in depth but distorted to give an apparent time structure due to the overlying channel. Depth conversion using raw moveout velocities results in the depth profile of the Latrobe surface shown in figure 11/10d. It should be noted that there is now an apparent structure in depth. It is imperative to avoid drilling such an anomaly in an expensive offshore exploration venture and so it is essential to correct for the distorting effect of the channel geometry before attempting to use moveout derived velocities for depth conversion.

As a practical example consider the seismic section A across the Kingfish Field referred to earlier in the text (figure 11/5). Figure 11/11 shows a normal moveout velocity profile for the top of the Latrobe Group as picked from

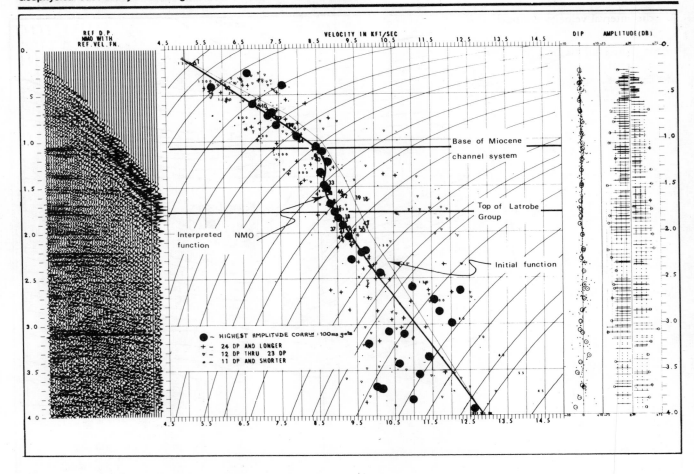

Figure 11/9: Typical scattergram in the Kingfish area .

the scattergrams along this line. The distorting effects of the channel geometry which we saw in the modelled situation are also expected here and they have been allowed for in the smooth line interpretation which shows the normal moveout velocities being too fast at the channel edge and too slow at the channel axis.

The smoothing of the raw moveout profiles to remove the estimated geometric effects of the channels is a key step in the velocity interpretation. The step is highly interpretive and requires considerable experience. Full use must be made at this stage of all local knowledge on velocity gradients, velocity contrasts, data quality, and channel base and fill geometry.

Once the profiles have been smoothed the intersections are checked for mis-ties and adjustments are made as necessary. When making these adjustments it should be kept in mind that seismic lines which are more nearly parallel to channel axes will suffer less from the distorting effects of channel irregularities. These lines are used as the key to eliminating or minimising the effects of ray path distortion on profiles which cut the channels at higher angles.

A map of smoothed moveout velocities is constructed after the smoothed profiles have been tied. Ideally, at any point on this map, the contours represent the moveout velocity which would be obtained for the top of the Latrobe Group at that location if there were 'layer-cake' geology with no channel problems. If done successfully the

smoothing process will have removed the distortion due to the outer rays of the common depth point gathers passing through a significantly different section to the more vertical rays.

The smoothed moveout velocity map must now be converted to an average velocity map by applying a percentage conversion factor. The conversion factors are determined at the well control by comparing the smoothed moveout velocity directly with the velocity obtained from the well survey. There is a danger however, that the smoothing of the velocity profiles is in error and this would obviously lead to an erroneous conversion factor at that well. To guard against this problem, a ray path modelling programme which uses the sonic log data is used to predict the normal moveout velocity and the conversion factor at each well. This computed value of the conversion factor is used to verify the smoothing of the profiles.

Away from the well control the prediction of conversion factors is more difficult and relies on an appreciation of the factors which cause the difference between the normal moveout velocities and the true velocities. Ray path model studies at the wells indicate that only three parameters significantly influence conversion factors. These are:

1. The effective offset (maximum source-detector separation after muting);

2. the depth to the top of the Latrobe Group; and

3. the interval velocity contrasts within the post-Latrobe section.

In the Gippsland Basin it is found that both the effective offset and the depth to the top of the Latrobe Group do not change enough over the area of a prospect to have a significant effect and this leaves the third parameter as the controlling influence in local conversion factor variations.

Experience shows that there is a bigger difference between the normal moveout velocities and the average velocities when the seismic rays travel through a thicker section of high velocity channel fill sediments. The reason for this is related to the greater refraction of the ray paths in the channelled areas, giving a bigger difference between the straight-line and the minimum travel-time paths. Consequently, channel base maps and channel fill isopachs are used to guide the trends of the conversion factor map between the well control. The conversion factor will decrease towards the channel axes where the average velocity is a smaller percentage of the moveout velocity. This effect is illustrated in figure 11/12 where conversion factors have been determined at eight wells in the Kingfish area and verified by ray path modelling. Comparison of this conversion factor map with channel base map in figure 11/6 shows the predicted decrease in conversion factors towards the channel axis and also the anticipated similarity in form between the two maps.

The conversion factor map shows a total variation of about 1.5 per cent across the Kingfish Field. Although this represents only about a 1.3 per cent variation in the average velocity it corresponds to a difference of about 100ft in depth. Such a change across the Kingfish Field could alter the estimated reserves by more than 100 million barrels and so it can be seen that the conversion factor variations can be highly significant.

At this stage in the interpretation procedure the conversion factor map and the smoothed moveout velocity map are combined to give an average velocity map to the top of the Latrobe Group (figure 11/13). This map is then used to produce a depth map to the top of the Latrobe Group from the seismic reflection times.

The velocity analysis technique which has been discussed in detail above was sufficient to define the size and shape of the Kingfish Field for development purposes. Subsequently, on the basis of this velocity interpretation technique and additional seismic data, the Kingfish-5 well was drilled to test a velocity dependent northeastern extension of the field. The top of the reservoir was encountered at a depth of 221ft below that predicted by the seismic interpretation and 35ft below the level of the oil-water contact within the field. Although this well did fail to encounter hydrocarbons, the revised structure map, based on this new well control and an updated velocity picture, showed an extension of the field to the northeast, although not as significant as predicted in the pre-drill analysis.

Later, the Kingfish-6 well was drilled one mile SSW of the Kingfish-5 well to further test the eastern extension of the field. This well also failed to encounter hydrocarbons and intersected the top of Latrobe reservoir 63ft deeper than predicted. The depth error in both of these wells was due to an error in predicting average velocity beneath the complex channelling of the Miocene section.

In the light of these disappointing well results it was obvious that a more accurate method of converting seismic

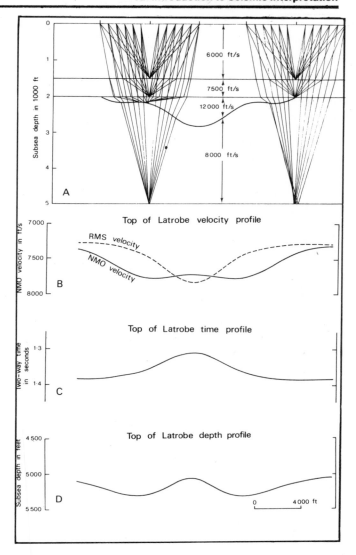

Figure 11/10: a) Raytracing model used to investigate the distortion of normal moveout hyperbolae beneath channels. **b)** Top of Latrobe velocity profile beneath the high velocity channel fill. **c)** Top of Latrobe time profile. **d)** Top of Latrobe 'depth' profile derived from the distorted moveout velocities.

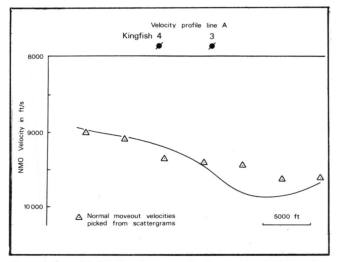

Figure 11/11: Profile showing normal moveout data points and a smooth interpretation in which the distortions due to the geometry of the base of the channel have been removed.

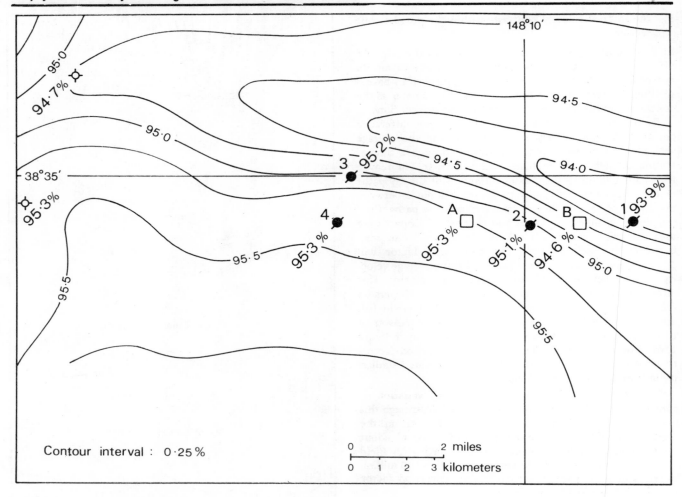

Figure 11/12: Conversion factor map for the top of the Latrobe Group.

reflection times to depths was needed in order to better define both the reserves of the field and the geometry of the flanks. The opportunity for this came in 1974—75 when a seismic survey was shot which gave considerably higher resolution seismic data and, in particular, better quality scattergrams than had previously been available. This improvement was achieved, in part by placing strict control on several factors affecting normal moveout determination. The gun to cable offset was carefully monitored and stringent signal/noise ratio specifications were prescribed. These factors plus the use of a 2msec sample rate and a relatively short group interval on the cable ensured good definition of the moveout curves.

Additional factors contributing to the improved data quality were the strict control of processing parameters and the monitoring of the data at several stages through the processing stream. Random errors in normal moveout velocity analysis, due to such effects as noise and multiples intersecting the primary events, were again minimised by use of the GSI '700 package' velocity analysis program. The new data gave a marked improvement in normal moveout velocity determinations, particularly from the shallow section. The increase in resolution also allowed a better definition of the complex channel geometries which are the major cause of moveout velocities being unreliable

indicators of true velocity. With these improvements it became feasible to make use of a method of depth conversion based on interval velocities.

11.4.3 The interval velocity approach to depth conversion at East Kingfish

At the eastern end of the field, Kingfish-5 encountered channel fill sediments with interval velocities as high as 14,150ft/s compared with the 9,400ft/s of the adjacent shales and marls. This velocity contrast of almost 5,000ft/s is more severe than seen in the central and western parts of the field and the time-depth conversion problems are correspondingly more complex.

A top of the Latrobe depth map prepared after completion of the Kingfish-6 well and produced from a one kilometre grid of 1974 high resolution data using the conventional conversion factor method showed two ridges running east-west along the crest of the structure (figure 11/14). The position of these ridges coincides with shoulders seen on the top of Latrobe reflection on the east Kingfish seismic sections (figure 11/16). These shoulders lie beneath the high velocity channel edges. It was difficult to explain these shoulders as real geological features and suspicion was thrown on the smoothing of the moveout velocity profiles. Consequently, in view of the improved data quality obtained in the 1974 shooting and the errors in depth prediction at Kingfish-5 and Kingfish-6, it was decided to apply a detailed interval velocity analysis over the eastern end of the field.

This interval velocity approach was a 'layer cake' method with the post-Latrobe section being divided into some fourteen distinct sedimentary layers and wedges using high resolution seismic data such as shown in figure 11/16. The digitised time horizons were then merged with a velocity file made up of interpreted scattergram data and the output was a set of Dix interval velocity profiles for each seismic line. These profiles were smoothed and tied with the cross lines. Once again, lines parallel to the channels were believed in preference to the cross lines where the geometric effects are the greatest. When all the horizons had been tied a series of interval velocity maps were made and a set of isopachs generated. Summing the isopachs produced a depth map to the top of the Latrobe reservoir. This map was slightly too deep because the interval velocities were derived from normal moveout velocities rather than rms velocities and a small correction factor was needed to tie the well control.

The depth structure map on the top of the reservoir produced by this interval velocity approach (figure 11/15) shows a smooth structure without the anomalous double ridges observed previously.

The depth picture obtained from the interval velocity technique was tested using a ray tracing program. Depth cross sections together with the smoothed interval velocities were input and the program traced the zero offset ray-paths. The first output was a set of simulated time sections which could be compared with the original data. The effects of migration due to the structure dip of the reflections and also due to refraction across dipping interfaces were apparent at this stage and the depth models were slightly modified to correct for the discrepancies.

The program was then used to simulate scattergrams along the seismic lines. These synthetic scattergrams and the derived normal moveout profiles to the top of the Latrobe Group compared favourably with the actual data and gave confidence to the depth interpretation.

In practice it often happens that the simulated time sections and scattergrams do not match the basic data sufficiently well to confirm the depth interpretation. In such cases, the parameters used to generate the depth models would be adjusted and the ray tracing analysis repeated until good agreement was obtained.

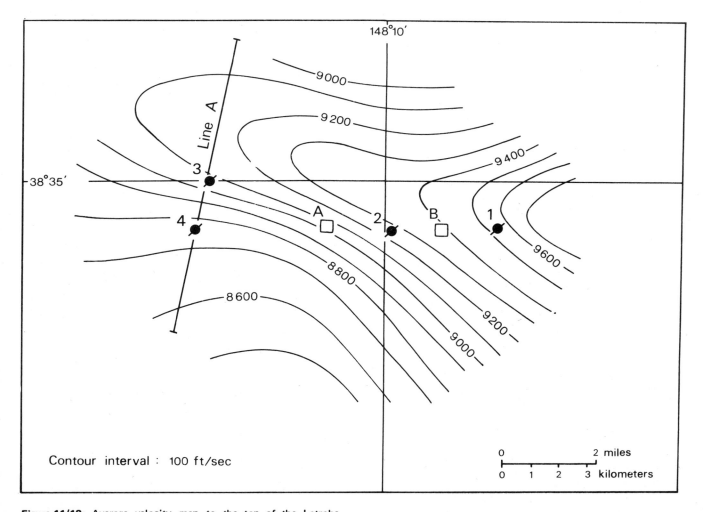

Figure 11/13: Average velocity map to the top of the Latrobe Group at Kingfish.

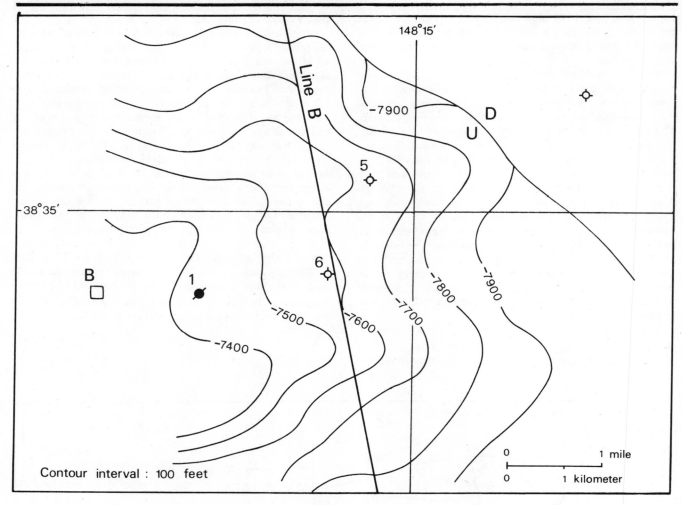

Figure 11/14: East Kingfish — Depth to the top of the Latrobe reservoir based on the conversion factor technique.

11.5 Summary

The rational exploration and development of the Gippsland Basin demanded that detailed velocity analyses be made over every prospect. The general method that has been adopted for deriving average velocities from seismic velocities is simple, with the most critical steps in the procedure being the smoothing of the moveout velocity profiles and the prediction of conversion factor variations. Both of these steps are highly interpretative and can only be carried out successfully with a sound knowledge of the local geology and an appreciation of the influence that the geology can have on seismic velocities. In particular this method has been applied in the Kingfish area where it has proved successful in defining the shape of the field sufficiently for proper development and reservoir management.

In areas with complex velocity problems but good seismic velocity data, such as the eastern part of the Kingfish Field, the interval velocity method can be used as an alternative method of depth conversion. This method has the added advantage that ray tracing programs can be used to check the results.

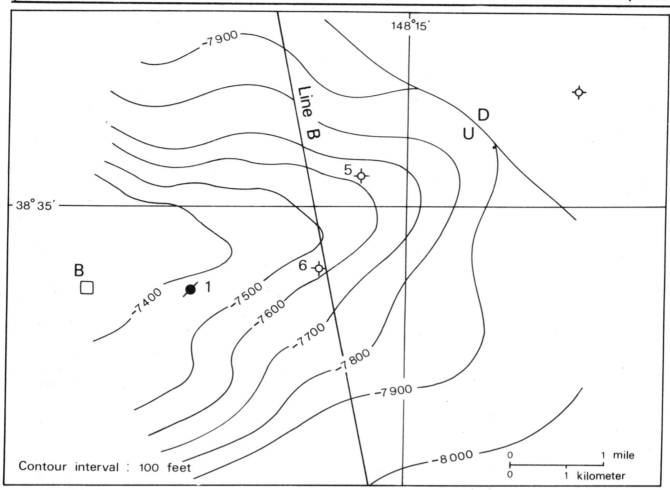

Figure 11/15: East Kingfish — Depth to the top of the Latrobe reservoir based on the interval velocity technique.

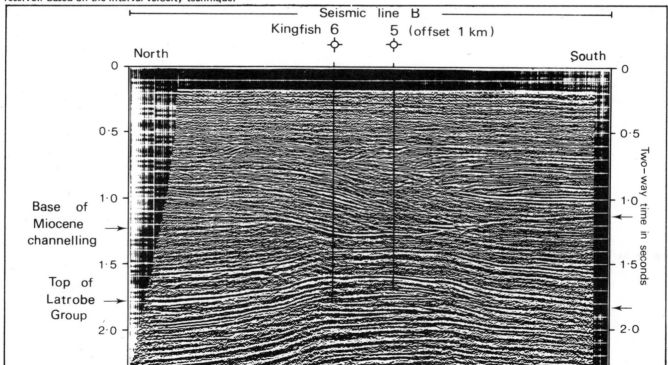

Figure 11/16: High resolution seismic line (line B in figures 11/14 and 11/15) across east Kingfish. GSI 1974 vintage survey.

12. THE HEWETT GAS FIELD CASE HISTORY

In 1959, at Groningen onshore in the northeast Netherlands, NAM, the Shell-Esso consortium, drilled the discovery well on what eventually proved to be a giant gas field with reserves of 1.642×10^{12} m^3 (58 T.c.f.) (see figure 12/1) The productive reservoir was in Rotliegendes (Permian) dune sandstones. Subsequent to the discovery, aeromagnetic surveys were conducted over the North Sea. When integrated with the regional geology, they confirmed predictions, as summarised by Wills,[*] of extension of Permian and younger basins westwards under the North Sea to the English mainland. (See figure 12/2).

In 1962, seismic reflection surveys commenced in the southern North Sea; the following year, three separate consortia, headed by the respective operators, Phillips Petroleum, Atlantic Refining Co. and Gulf each commissioned three separate surveys on a ½° x 1° latitude and longitude grid. They were planned to interlock with each other and after trading each other's data, provided a grid of 20 km square, ie approximately subdivided in 10′ x 20′ areas. A fourth consortium, led by Signal Oil[†], joined this arrangement, shooting infill lines in the southern part of the survey area; the final programme grid spacing was on a 5′ x 10′ grid, or approximately 10 km square (figure 12/3). The 13,600 line km were obtained over an area of 114,000 km^2. The data were acquired by three different contractors, Geophysical Service International, Seismograph Service Limited and Robert H. Ray[‡]. Single-fold split-spread profiles were obtained utilising a two-boat operation and dynamite shooting. Navigation and position fixing was by Decca Sea Search involving the use of a master and two slave stations.

From the outset, interpretation revealed the existence of large structures, trending mainly NW-SE, approximately parallel with an ancient high to the southwest, the London-Brabant massif. Correlation of the seismic with onshore well information in the UK, Netherlands and Germany, combined with evidence of extra strong reflectivities and strong absorption led to identification of Permian salt structures. Although these were not as intensely deformed as were the well-documented salt domes and diapirs of Germany, they still showed pronounced relief; in response to rejuvenation of normal faults in the underlying Permian and Carboniferous rocks, the highly plastic halite had flowed, creating folds and salt collapse structures in the overlying Triassic, Jurassic and Cretaceous strata.

From the 1964 Zechstein time structure map shown in figure 12/4 and the accompanying regional geological section, (see figure 12/5) Phillips Petroleum identified one large structure, 15 miles off the coast of East Anglia. In the UK 1964 First Round of Offshore Production Licence awards, the Phillips Consortium were successful in their application for blocks 48/30 and 52/5, over the southeast part of the structure; the Arpet Group became the successful licensees in blocks 48/28 and 29 over the northwest half of the structure. Dip and strike orientated surveys were commissioned in 1964 and 1965. By this time multifold or common depth point shooting had 'arrived' in the North

---

[*] L.J. Wills *Palaegeographical Atlas*
[†] Subsequently merged into Burmah Oil, the North Sea division of which was later incorporated into BNOC Development Ltd.
[‡] Now known as Geosource Petty-Ray.

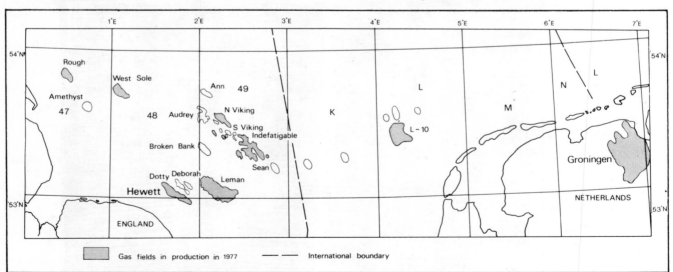

Figure 12/1: Gas fields of the southern North Sea and the Netherlands.

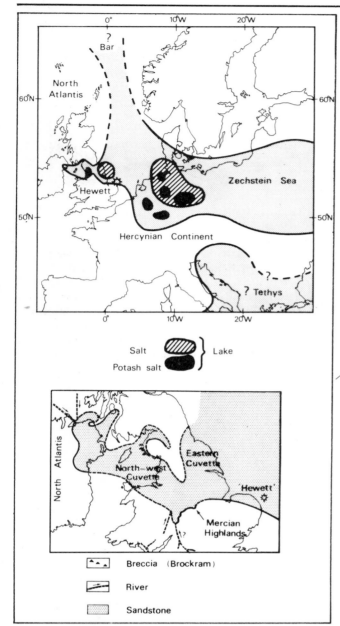

Figure 12/2: Palaeogeography of the Upper Palaeozoic/Lower Mesozoic: *ia)* **(top)** Late Permian: the Zechstein Sea extended from E. Europe to Ireland. Fluctuations of evaporite cycles led to precipitation of large deposits of halite and/or potash salts. The Hewett sands were deposited immediately on top of the Zechstein presumably after some rejuvenation of the nearby London-Brabant massif. *b)* **(bottom)** Lower Trias (Middle Bunter): Lower Bunter sands ('Hewett' sands) have limited distribution, near highland areas; Middle Bunter sands (the 'Bunter' of Hewett) were deposited over a wide area in large delta fans. *(after Wills, 1962).*

Sea and most lines were recorded and processed 3- or 6-fold with a welcome increase in quality. Based on the interpretation of these, Arpet spudded a well on the structure and in late 1966 penetrated significant gas zones in two separate Triassic sandstone reservoirs, with a lesser show in the Permian Hauptdolomit formation. A stepout well, 48/29-2, 2.4 km to the SSE, found gas in these same three zones, and immediately following this, the Phillips Group made a discovery 13 km to the southeast of the original 48/29-1 well confirming the large areal extent of the reservoirs. The Hewett Field had been discovered.

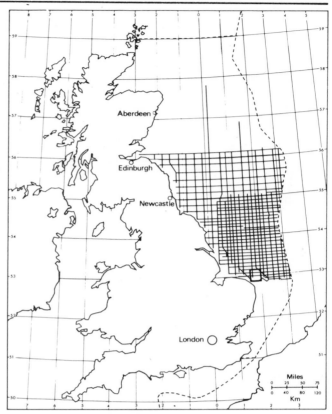

Figure 12/3: Initial seismic reconnaissance of the UK North Sea: the 1963 Phillips/Arpet/Gulf/Signal Group Surveys. The dense grid spacing is 10km square: the rest is 20km square. The map area of figure 12/4 is highlighted *(after Phillips Petroleum Expl. UK Ltd).*

12.1 Seismic interpretation

Several hundred miles of line were shot over the structure between 1963 and 1968. Initially, data were obtained using a two-boat operation, with dynamite charges up to 110 kg, split-spread cables up to 1,200 metres either side of the shotpoint, 20-25 geophones per station and analogue recording. Gradually, single-fold was superseded by 3-fold and 6-fold stacking, and analogue instruments gave way to digital; new sources evolved improving on production efficiency, signal enhancement and environmental acceptability. The oldest data suffered from very poor normal moveout control over the first 0.5 seconds and as a result the stacking is not the best. Pre-deconvolution data suffer very badly from water bottom reverberations. This shallow-water area (40-20m over the crest of the structure), due to its location near the coastline, the confluence of the southern North Sea with the English Channel, and the Humber, Wash and Thames estuaries, is subject to strong tidal variations. This factor, combined with the limited resolution of the North Sea first phase position-fixing instrumentation, led to navigational and surveying errors which resulted in poor stacking and line mis-ties.

Four lines, two dip-orientated NE-SW, and two strike orientated lines, NW-SE, have been chosen to highlight the seismic interpretation of the Hewett structure; together, they form a closed loop. They are of variable quality, not only because they were shot and processed at different periods during a rapidly changing seismic technology, but also because of their particular orientation relative to the complex-faulted Hewett structure. These seismic sections are reproduced in figures 12/6, 7, 8 and 9.

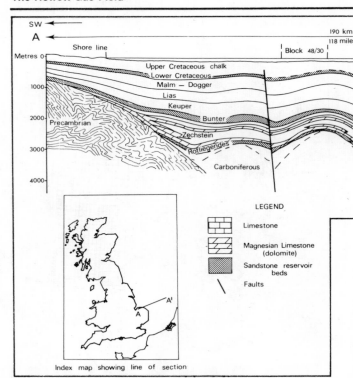

Figure 12/5: NE-SW geological section, offshore East Anglia as prepared in 1964, *(after Phillips Petroleum Expl. UK Ltd).*

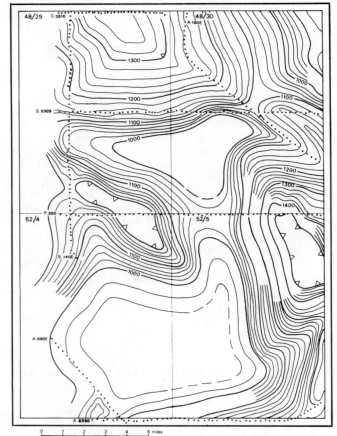

Figure 12/4: Zechstein time structure map from reconnaissance data prepared for the 1964 licence application. Despite the loose (10km) control, the existence of broad structural highs were outlined by the interpretation. Of these blocks, two each were awarded to the Phillips and Arpet groups. *(after Phillips Petroleum Expl. UK Ltd).*

Of several original reflection events, we will confine ourselves to the three most important:*.

1. Zechstein: This is the strongest event on the sections. A strong velocity contrast is obtained between the Hewett sandstone with a velocity of around 12,000 ft/s and evaporites, salts and some shales ranging down from 16,500 ft/s through 14,000 ft/s to 12-14,000 ft/s, respectively. This event is one cycle above the Z3 or Hauptanhydrit event which is the deepest continuous reflector available for mapping the underlying Rotliegendes reservoirs of the southern North Sea. (The Zechstein halite acts as an absorber of seismic energy and underlying events can only be mapped by undershooting or refraction).

2. Bunter Sandstone: This varies in quality but is generally continuous. A velocity reversal obtains between the overlying compacted shales and the partly consolidated highly porous sandstone, when gas-filled. When tight or water-filled, the velocity contrast diminishes and a normal or weak event is obtained.

3. Keuper: This is an event near the top of the Triassic. It is generally continuous and of good quality throughout the area, except on the older data: it comes in as early as 0.4 seconds and is often buried in noise or poorly NMO-corrected traces. Reflection generation is from halite beds overlain by mudstones.

12.2 Seismic sections

Four interpreted seismic sections are shown in figures 12/6, 7, 8 and 9, locations of seismic lines are shown in figures 12/10 and 12/11.

Although the Keuper reflection is an important event, being the probably Early Cimmerian unconformity and consequently being important in terms of identifying the structural growth history, on most of these sections, the reflection is NR and therefore not identified: for identification see the recent sections in figures 12/17a and 12/17b.

HN (Ray, 1965, 6-fold, dynamite, 1680 m. split-spread cable, 70 m. offset, no deconvolution).

This NE-SW dip line is located 1 km. northwest of the Phillips discovery well 52/5-1, on the apex of both the Hewett and Bunter sandstone reservoirs. The major bounding faults are seen at the Bunter level at SPs 1½ and 7.

---

* See figure 12/15 formation velocities.

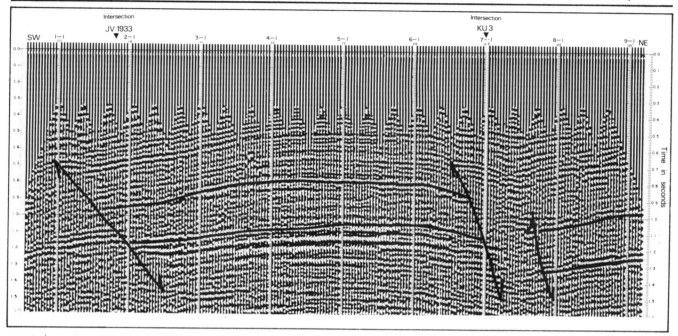

Figure 12/6: Seismic section of line HN, dynamite, 6-fold stack without deconvolution *(Ray, 1965)*.

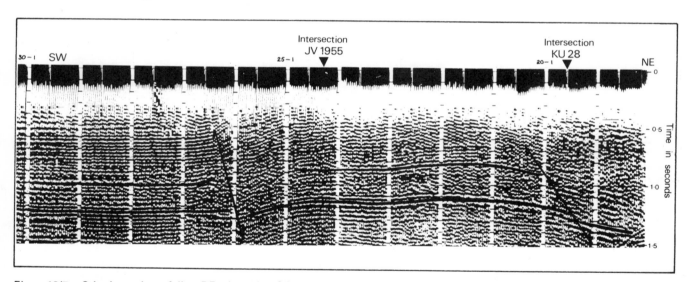

Figure 12/7: Seismic section of line DF, dynamite, 6-fold stack without deconvolution *(Ray, 1965)*.

Correlation of both the Bunter and Zechstein reflections across the southwestern fault cannot be made. Correlation over the northwestern fault is reliable; because of the uncertainty of fault alignment, the two faults shown here are shown as one broad fault zone on the respective maps. The ties with line KU and JV are good at the Zechstein level. At the Bunter level, the tie with JV is good but the tie with KU is impossible, due to its occurrence at the coincidence with the major north western bounding fault. Reflection quality is good and the picks are reliable over the main structure. Note the occurrence at the 0.8 to 0.9 second levels, between SPs 3½ and 6, of a flat, slightly dipping reflector. As suggested by Cumming and Wyndham, its contrast with the anticlinal curvature of the structure and its flatness suggest it is a gas/water contact.

DF (Ray, 1965, 6-fold, dynamite, 840 m. split-spread cable, 35 m. offset, no deconvolution).

This dip line is located to the southeast of the structure, as it plunges in that direction. The major northeastern bounding fault is again apparent at SP 20, at Bunter level. The major bounding fault in the southwest, which was not seen on line HN, occurs at SP 26. The steep dips into the SW bounding fault suggest, as nearby lines show, that a fault occurs around SP 25 throwing down to the southwest. Ties with line JV are good; the ties with line KU are not effected due to the proximity of the major northeastern bounding fault. Reflection quality is fair, and the picks are reliable at both the Bunter and Zechstein levels, especially over the crest of the main structure.

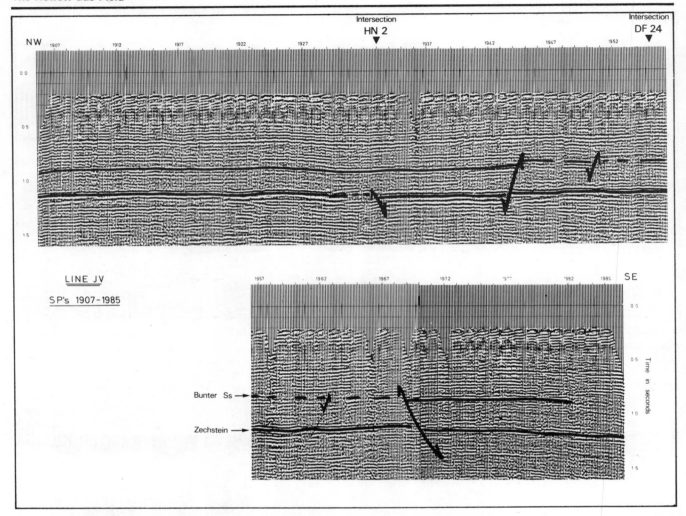

Figure 12/8: Seismic section of line JV, dynamite, 3-fold stack without deconvolution *(SSL, 1967).*

JV (SSL, 1967, 3-fold, dynamite, 1200 m. split-spread cable:

Being a strike line, the structural alignment is relatively uninteresting, running as it does along the southwest flank of the structure, close to the southwestern bounding faults. Ties with lines HN and DF are good at both levels. Several faults are apparent at both the Bunter and Zechstein levels: they are offshoots from the main southwestern bounding fault. Picking is not so reliable as on the dip lines, but the common water bottom reverberations do not help. The reflection strength of the Bunter is weak or irregular over most of the section; however in the southwest, it is particularly strong between SPs 1970 and 1980. This is the part of the line closest to the Bunter gas/water interface (see figure 12/12), and the strength of the reflection may be due to 'gas effect', as described later in the discussion of hydrocarbon indicators.

KU (Western, 1968, 12-fold, Aquapulse (4 guns; 4 pops per shotpoint, summed), 1200 m. cable, deconvolution before and after stack):

This NW-SE strike line ties lines DF and HN at their northeastern extremities. Data quality is quite good; however, as this parallels the NW-SE northeasterly bounding fault, diffractions and energy loss make reflection continuity across the faults doubtful, and the line ties difficult. Only one tie is made, with line HN at the Zechstein level, on the upthrown side of the major fault. This, in fact, is the only portion of the main Hewett structure shown on this section. Identification of the Bunter on the up-thrown side of the fault is not possible, due to loss of fold and fault effects. Identification of the Bunter and Zechstein on the downthrown side of the fault is by character correlation.

The picking and mapping of lines HN, DF, JV and KU is shown on the two seismic time structure maps, 'Bunter Sandstone' (figure 12/10). and 'Zechstein' (figure 12/11). The remainder of the contouring on these two maps is abstracted from Phillips in-house maps (based on interpretation of the seismic sections shown in faint dashed lines in figures 12/10 and 12/11). Although the mapping

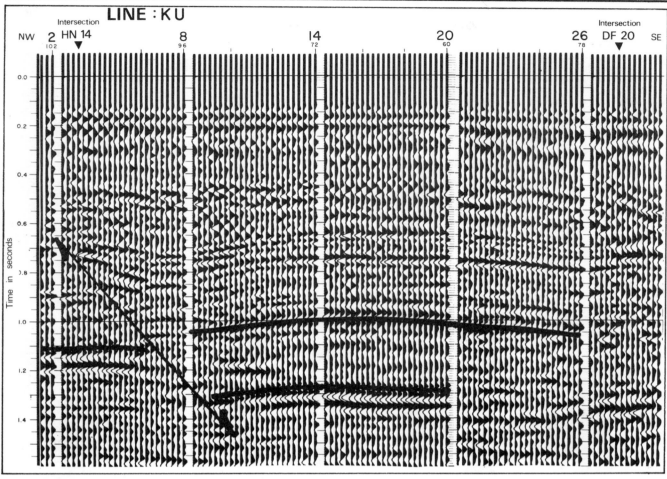

Figure 12/9: Seismic section of line KU, Aquapulse, 12-fold stack with deconvolution before and after stack *(Western, 1968).*

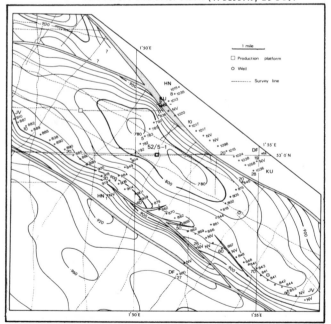

Figure 12/10: Bunter Sandstone time structure map over the SE area of the Hewett Field: SP locations and two-way reflection time values for seismic lines HN, DF, KU and JV are shown; faint dashed lines indicate the location of other lines used for the complete interpretation of this horizon (note the irregularity of many of these lines due to tidal influences and the limited resolution of initial position-fixing systems).*(After Phillips Petroleum Expl. UK Ltd).*

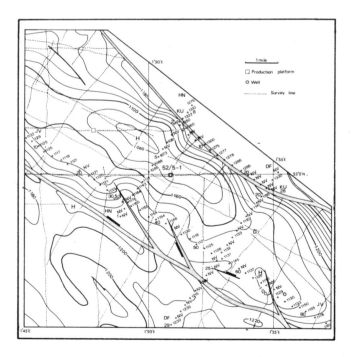

Figure 12/11: Zechstein time structure map, with location of seismic lines HN, DF, JV and KU. The Zechstein immediately underlies the Hewett Sandstone *(after Phillips Petroleum Expl. UK Ltd).*

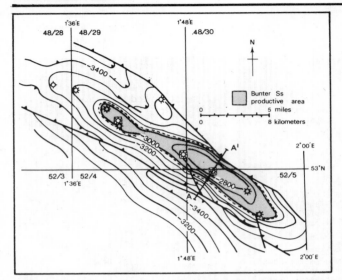

Figure 12/12: Bunter Sandstone depth map *(after Cumming and Wyndham, 1975).*

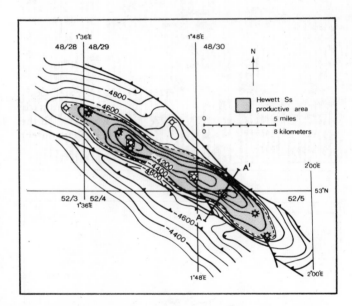

Figure 12/13: Hewett Sandstone depth map *(after Cumming and Wyndham, 1975).*

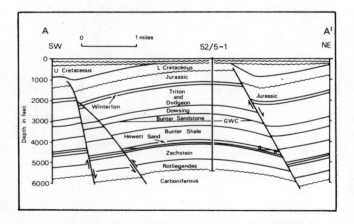

Figure 12/14: Geological dip section A-A', through Phillips' discovery well 52/5-1. Line of section is shown in figures 12/12 and 12/13 *(after Cumming and Wyndham, 1975).*

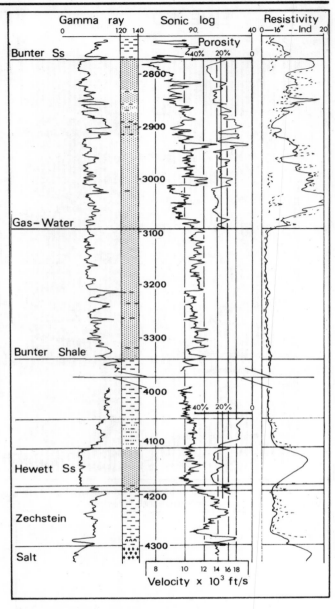

Figure 12/15: Wireline logs of type section. Velocity scales have been superimposed on the sonic log *(after Cumming and Wyndham, 1975).*

of these reflections suggest a simple folded structure, the complexity of the bounding fault relationships is well illustrated by close analysis of these seismic sections. This is particularly so in the southwest, where matching of the fault strikes at the Bunter and Zechstein levels is difficult. At the Bunter level, the main offset from the southwestern bounding fault throws down to the southwest on line DF, creating a small graben, whereas on line HN, it throws down to the northeast, creating a southwesterly dipping block between itself and the main southwestern bounding fault.

When integrated with the valuable well velocity and gas/water interface data, the seismic maps were converted to depth and isopach maps which provided the base for reservoir development studies. The final structure maps for both reservoirs are shown in figures 12/12 and 12/13. A geological dip-oriented section of the reservoir, through the Phillips discovery well 52/5-1 and parallel with seismic line HN, is illustrated in figure 12/14. Typical sonic and other wire-line logs are compared in figure 12/15.

The Hewett Gas Field is unique in the southern North Sea in being Triassic; the underlying Rotliegendes which is the reservoir for all other southern North Sea fields, is thought to lack a suitable seal, as the faulting atypically extends up through the Zechstein salt. These same faults probably provided the vertical migration path for the gas which is thought to have the same source as the Rotliegendes fields, i.e. from Carboniferous coal beds; however, it should be noted that the gases in the two reservoirs are different in their chemical composition, the Bunter alone having hydrogen sulphide, and much higher carbon dioxide and nitrogen (Cumming & Wyndham). Shales and mudstones, with anhydrite provide the seals for both the Bunter Sandstone and Hewett Sandstone reservoirs.

Seismic surveys and a total of seven exploration wells established the field outlines as shown in figures 12/16. The lower reservoir extends over 27 km. NW/SE, is 4.34 km. in maximum width, and has an area of approximately 90 km^2 (35 square miles). Total recoverable reserves are in the neighbourhood of 85 x 10^{9}m^3 (3 T.c.f.).

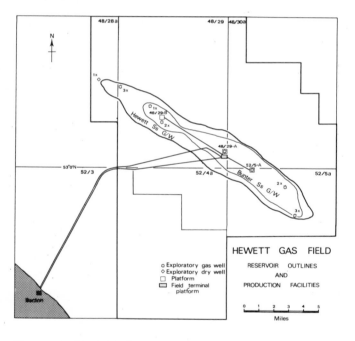

Figure 12/16: The Hewett Gas Field. The limits of the two Triassic Sandstone reservoirs are shown: the field covers an area of 90 sq. km. and is produced from three production platforms *(after Phillips Petroleum Expl. UK Ltd).*

A unitisation agreement was signed between the Arpet and Phillips Groups in April 1969 and production from the first of three platforms commenced in July of that year. The combined group were successful in their application for Block 52/4 in the 1970 third round of bidding: after the statutory 50 per cent relinquishments, made on the sixth anniversary of the initial awards in the first and third rounds, the field is now completely contained by the part blocks 48/28a, 29a and 30a, 52/4a and 5a, all of which are licensed to the combined Phillips and/or Arpet Group.

The complete list of companies is as follows:
Phillips Petroleum Exploration UK Ltd. (Operator for
  Phillips Group and Hewett Unit).
  Agip (UK) Ltd.
  Fina Exploration Ltd.
  Century Power and Light Ltd.
  Halkyn District United Mines Ltd.
  Oil Exploration Ltd.
  Plascom (1909) Ltd.
Arpet Petroleum Ltd. (Operator for Arpet Group).
  British Sun Oil Company Ltd.
  North Sea Exploitation and Research Company Ltd.
  Superior Oil (UK) Ltd.
  Canadian Superior Oil (UK) Ltd.
  Sinclair (UK) Oil Company Ltd.

12.3 Hydrocarbon indicators

Subsequent to reservoir studies and the completion of development drilling, it was decided to utilise newly developing techniques of direct hydrocarbon detection. In 1974 a short survey was run over the field using the best acquisition and processing parameters available. Two of the lines are shown in figure 12/17a and 12/17b:

Figure 12/17b: The NE/SW dip line parallels the line HN, discussed earlier. It also passes very close to the well 52/5-1 which was drilled near the crest of both reservoirs. Comparison of the two lines serves to illustrate the dramatic advance in acquisition and processing. Concordant with the increase in quality, there is an increase in frequency content and it is now possible to identify the Hewett Sandstone reflections. In addition, the Zechstein events have acquired high resolution character; what was thought on the early data to be a 0.05 seconds/20 Hz wavelet is now a 0.05 seconds/40 Hz doublet with the null amplitude zone almost certainly defining the absorptive Zechstein halite layer. The faults are now much easier to define. Two hydrocarbon indicator criteria are quite noticeable: a strong black peak, indicating a negative accoustic impedance appears at the top of the gas-laden Bunter Sandstone; as was seen on line HN, but here with much better resolution there is also what appears to be a dipping gas water contact underneath the Bunter Sandstone top. The dip in the contact is probably due to a velocity gradient which follows the dip of the structure.

Figure 12/17a: This strike line follows the crest of the structure from the northwest of 48/29-2 to southeast of 52/5-3. It is most notable structurally that a saddle separates the major part of the Hewett structure, in the southwest, from the two 'satellite' extensions in the northwest. The superb quality of this section almost makes the necessity for true amplitude sections redundant. Several striking hydrocarbon indicator criteria are apparent:

1) a flat-lying event below the Bunter Sandstone reflector, between shot-points 145-220.
2) a negative accoustic impedance peak for the Bunter Sandstone reflector, between shot-points 1-240.
3) a second leg 'bright spot'/thin bed tuning as the negative accoustic impedance peak and the flat-lying event converge, viz, at shot-points 35-63 and 90-110.

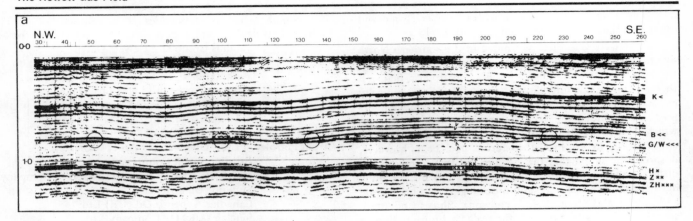

Figure 12/17: Seismic sections derived from 1974 vintage surveys. *a)* Strike section along axial crest of the field. Most noteable character effects are gas-induced negative reflectivity, amplitude diminishing off-reservoir, gas-water contact reflection and quarter-wavelength 'tuning' which is circled in the figure.

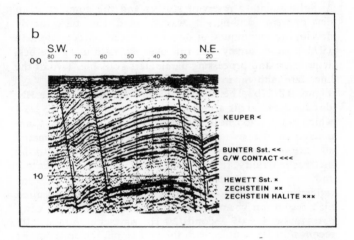

b) Dip section near 52/5-1. This is a remarkable improvement on the data from line HN which is nearby. The severe faulting at the margins obscures some hydrocarbon criteria, however the gas-water contact in the Bunter Sandstone is apparent, and there is even a suggestion of one in the Hewett Sandstone *(both sections after Phillips Petroleum Expl. UK Ltd).*

Reflection coefficients and amplitude response analysis

Reflection coefficient analysis of the hydrocarbon indicators mentioned above is hampered by the difficulty in obtaining the sub-surface gas velocities. For the Bunter reservoir, the following averaged (and rounded) parameters have been determined.

V_{bm}, velocity of Bunter Sandstone matrix = 18,000 ft/s
ρ_{bm}, density of Bunter Sandstone matrix = 2.65 g/cc
V_w, velocity of reservoir water = 5,000 ft/s
V_{sh}, velocity of overlying shales = 10,500 ft/s
ρ_{sh}, density of overlying shales = 2.30 g/cc
ϕ, porosity of the reservoir = 0.257
S_w, water saturation of reservoir = 0.137
V_g, velocity of Bunter gas = 1,400 ft/s
ρ_g, density of Bunter gas = 0.07 g/cc

Using the Wyllie Time Average Equation (chapter 1), the velocity of the gas saturated Bunter Sandstone

$$V_{bg} = \text{reciprocal of } \phi \frac{S_w}{V_w} + \frac{\phi (1 - S_w)}{V_g} + \frac{(1 - \phi)}{V_{bm}}$$

$$= 4,840 \text{ ft/s}$$

This is considerably below the lowest sonic-derived values, which averaged 10,070 ft/s from four wells, and which were probably highly affected by mud invasion. For further use, a more compatible value has been extracted from a recent paper on fluid saturated reservoir velocities. It more closely matches the lowest values seen on the sonic logs; it is $V_{bg} = 8,500$ ft/s (Gregory 1976).

For V_{bw}, velocity of Bunter sandstone, water-saturated, the Wyllie equation can be used:

$$\frac{1}{V_{bw}} = \frac{\phi}{V_w} + \frac{(1 - \phi)}{V_{bm}}$$

$$= (0.257/5,000) + (0.743/18,000);$$

$V_{bw} = 10,800$ ft/s (Average sonic velocities of four wells was 11,330 ft/s, but again they are probably affected by mud-invasion).

For ρ_{bw}, reservoir density, water-saturated.
$$\rho_{bw} = \phi + (1 - \phi) \rho_{bm}$$
$$= 0.257 + (0.743 \times 2.65)$$
$$= 2.23 \text{ g/cc}$$
Similarly $\rho_{bg} = \phi S_w + \phi(1 - S_w) \rho_g + (1 - \phi) \rho_{bm}$
$$= (0.257 \times 0.137) + (0.257 \times 0.863 + 0.07) + (0.743 \times 2.65)$$
$$= 2.02 \text{ g/cc (Average log densities of}$$
four wells was 2.08, again probably too high because of mud effect).

The reflection coefficient for the top of the Bunter gas reservoir,
$$RF_{bg} = (V_{bg}\rho_{bg} - V_{sh}\rho_{sh})/(V_{bg}\rho_{bg} + V_{sh}\rho_{sh})$$
$$= ((85 \times 2.02) - (105 \times 2.30))/((85 \times 2.02) + (105 \times 2.30))$$
$$= -0.1689,$$

The reflection coefficient for the gas/water interface,

$$RF_{bw} = (V_{bw}\rho_{bw} - V_{bg}\rho_{bg})/(V_{bw}\rho_{bw} + V_{bg}\rho_{bg})$$
$$= ((108 \times 2.23) - (85 \times 2.02))/((108 \times 2.23) + (85 \times 2.02))$$
$$= 0.1676,$$

The values are rather higher than predicted by Gregory, but approximate the results of Toksoz *et al.*

The reflection coefficient for the off-structure top of Bunter, wet,

$$RF_{sh/bw} = (V_{bw}\rho_{bw} - V_{sh}\rho_{sh})/(V_{bw}\rho_{bw} + V_{sh}\rho_{sh})$$
$$= ((108 \times 2.23) - (105 \times 2.30))/((108 \times 2.23) + (105 \times 2.30))$$
$$= -0.0014$$

Knowing these reflection coefficients we can now draw several conclusions about the meaning of the different character events in figures 12/17a and 17b.

1) Flat-lying events immediately below the Bunter top are gas/water interfaces, reflection strength is moderate, but the events are distinctive because of their intrinsic horizontal aspect where other dips are well developed.

2) The negative accoustic impedance for the top of the pay zone is confirmed. Reflection strength is just moderate.

3) The near-zero acoustic impedance calculation for the top of a wet Bunter reservoir suggests that 'dim spots' will be produced from the edges of the gas reservoir, and down-dip. This is exactly what we see from SP 240 and south-eastwards in figure 12/17a.

4) With equal and opposite coefficients, the top pay zone and gas/water reflections are ripe for exhibiting tuning or bright spots as the pay thickness tends towards the $\lambda/4$ (one-way time; $\lambda/2$, two-way time) relationship. By comparison with the sections, a 0.040 second minimum phase Ricker wavelet seems the most likely to utilise in studying reflection response; with the above velocity $V_{bg} = 8,500$ ft/s, the $\lambda/4$ (one-way time) tuning occurs at 0.010 sec. or 85 feet. By coincidence, the pay thickness at 48/29-2 is 85 feet and if one looks at close-by SP43, in figure 12/17a we do get tuning and a bright spot, extending in fact to SP 62; this and three other similar effects are circled at SPs 98, 134, and 224. Apart from a good Bunter gas/water contact reflection, figure 12/17b fails to show other hydrocarbon criteria due to masking by faulting and diffractions.

A plot of Bunter sandstone amplitude versus time response is shown in figure 12/18. It was derived as follows:

A minimum phase Ricker wavelet of length 0.065 second and effective wavelength of 0.040 second (effective frequency 25 Hz) was used for the top of the pay sand response. By convention it is shown as a peak (negative reflection coefficient). For the gas/water response, the same wavelet was inverted and amplitude modulated to compensate for the partial attenuation of the wavelet at the preceding (gas sand) interface. Starting at 0.004 seconds wavelet separation (17 feet) with near maximum interference and near-zero amplitude response, the wavelets were separated in increments of 0.008 seconds (two-way time 34 feet), and their resultant responses plotted. The

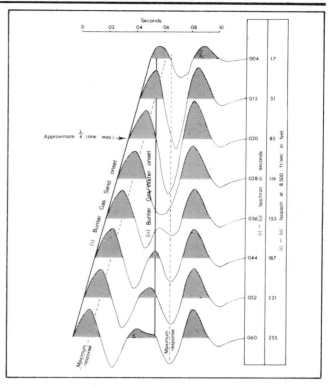

Figure 12/18: Bunter Sandstone wavelet response. Varying the gas pay thickness results in 'tuning' at 85' (0.020 second separation) and separation of the shale/gas and gas/water wavelets at 145 feet.

$\lambda/4$ wavelength (one-way time) reinforcement or tuning is readily apparent at 0.020 second separation, which is equivalent to a pay thickness of 85 feet. (Note that a reverse polarity presentation would give a very strong 'dim-spot'). Separation of the wavelets occurs around 0.034 seconds or 145 feet and is complete for a gas sand thickness of 170-264 feet (0.040 to 0.062 seconds, two-way time delay). The maximum Bunter gross gas pay thickness is 323', recorded at the Phillips discovery well 52/5-1. Given enough seismic data of the quality exhibited in figure 12/17a, and using the criteria developed here, the construction of accurate gas pay maps for the Bunter reservoir should be straightforward.

In this chapter we have used the Hewett gas field case history to briefly trace, over about 15 years, the development of seismic acquisition and interpretation techniques. This history commenced with the early phase of exploration for hydrocarbons in the southern North Sea in 1962, through the discovery phase to that of detailed reservoir studies and the use of modern direct hydrocarbon indicator techniques. As has been discussed in earlier chapters, modern exploration can benefit from what is now considered to be advanced technology even during the earliest exploration phase, but any case history can only relate those events which have occurred over a limited span of years, though in the oil industry a decade is quite a long period. Seismic techniques which were being heralded as revolutionary in the early 1960s are now obsolescent and surely the techniques of the late 1980s will have advanced far beyond those now in current use. Such advances cannot be accurately predicted, but hopefully they will allow the geophysicist to discover and contribute to the exploitation of hydrocarbon reserves which can now be regarded only as unquantified and ill-defined potential resources.

References

A.D. Cumming and C.L. Wyndham, 'The geology and development of the Hewett gas field, in A.W. Woodland (ed) *Petroleum and the Continental Shelf of North West Europe,* Vol. 1 Geology, (Applied Science Publishers, Barking, England, 1975).

A.R. Gregory, 'Fluid saturation effects on dynamic elastic properties of sedimentary rocks'. *Geophysics,* 41(1976) No. 5, pp. 895–921.

L.J. Wills, *A Palaeogeographical atlas of the British Isles and adjacent parts of Europe.* (Blackie and Son Limited, London, 1951).

M.N. Toksoz, C.H. Cheng and A. Timur, 'Fluid saturation effects on dynamic elastic properties of sedimentary rocks'. *Geophysics*, 41(1976) No.4, pp.621-645.

M.R.J. Wyllie, A.R. Gregory and L.W. Gardner. Elastic wave velocities in hetrogeneous porous media. *Geophysics*, 21(1956) 41-70.

Acknowledgement
Use of seismic sections, maps and other illustrations kindly permitted by the Hewett Operator, Phillips Petroleum, and its partners (as listed previously). Various Phillips and Arpet staff are to be thanked for the use of their work, Oil Exploration (Holdings) Ltd. for its encouragement, and Dr. T.C. Richards for personal communication on some of his work on reflection amplitudes. Unless otherwise stated, all seismic interpretive conclusions are the author's.

APPENDIX 1 ELEMENTARY SIGNAL PROCESSING THEORY

This appendix includes a mathematical treatment of signal processing which is presented with the aim of supplementing the theoretical treatment of this subject in earlier chapters. Though not essential to the main content of the book, the processes described are important, and the non-mathematical reader may still find gain in studying the illustrations without attempting a full appreciation of the mathematical arguments.

Fourier analysis

A concept of fundamental importance to seismic processing is that the signal can be represented by summing a series of sinusoidal oscillations with the correct amplitudes and phases.

Consider first a function $f(x)$ which is defined between $x = -\pi$ and $x = \pi$, and initially assume this function to be odd, so that $f(x) = -f(x)$. Then, under conditions satisfied by all functions of physical interest, $f(x)$ may be represented as the sum of a series such that:

$$f(x) = \sum_{n=1}^{\infty} a_n \cdot \sin nx$$

where the coefficients a_n are given by

$$a_n = \frac{2}{\pi} \int_0^{\pi} f(\xi) \cdot \sin n\xi \cdot d\xi.$$

This result may be proved by considering the integral (for integral values of m).

$$\int_0^{\pi} f(x) \cdot \sin mx \cdot dx = \sum_{n=1}^{\infty} \int_0^{\pi} a_n \cdot \sin nx \cdot \sin mx \cdot dx.$$

But $\int_0^{\pi} \sin nx \cdot \sin mx \cdot dx =$

$$= \frac{1}{2} \left[\int_0^{\pi} \cos(n-m)x \cdot dx - \int_0^{\pi} \cos(n+m)x \cdot dx \right]$$

$$= \frac{1}{2} \left[\frac{\sin(n-m)x}{(n-m)} - \frac{\sin(n+m)x}{(n+m)} \right]_0^{\pi}$$

$$= 0, \text{ for } n \neq m.$$

If $n = m$, $\int_0^{\pi} \sin^2 nx \cdot dx = \frac{1}{2} \cdot \int_0^{\pi} (1 - \cos 2nx) \cdot dx = \frac{\pi}{2}.$

Thus $\int_0^{\pi} f(x) \cdot \sin mx \cdot dx = \frac{\pi}{2} \cdot a_m$, which is the required result.

As an example of the analysis, consider the function $f(x) = 1$ for $0 < x < \pi$ and $f(x) = -1$ for $-\pi < x < 0$. Then the a_n are given by:

$$a_n = \frac{2}{\pi} \cdot \int_0^{\pi} \sin n\xi \cdot d\xi = \frac{-2}{n\pi} \left[\cos n\xi \right]_0^{\pi} = \frac{4}{n\pi} \text{ for } n \text{ odd,}$$
$$= 0 \text{ for } n \text{ even.}$$

An example of the way in which a series of sinusoidal terms sums to give a required function is shown in figure 1/1. We can visualise the frequency content of such a signal (that is, the amplitude of the a_n corresponding to different frequencies) if we plot the graph shown in figure 1/2, which is called the amplitude spectrum.

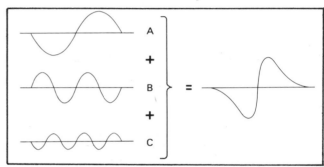

Figure 1/1: Fourier decomposition of a simple function.

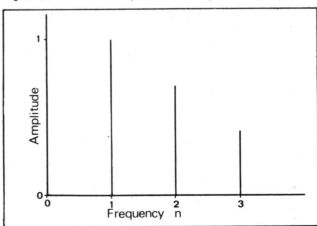

Figure 1/2: Amplitude spectrum of the function shown in figure 1/1.

If the function $f(x)$ is not an odd function, it may be expanded into a similar series, but both cosine and sine terms must be used:

$$f(x) = \sum_{n=1}^{\infty} a_n \sin nx + \tfrac{1}{2}b_0 + \sum_{n=1}^{\infty} b_n \cos nx,$$

where $\quad a_n = \frac{1}{\pi} \int_{-\pi}^{\pi} f(\xi) \sin n\xi \cdot d\xi,$

and $\quad b_n = \frac{1}{\pi} \int_{-\pi}^{\pi} f(\xi) \cos n\xi \cdot d\xi.$

In this case, we can talk both of the amplitude of a particular frequency ($\sqrt{a_n{}^2 + b_n{}^2}$) and of its phase (arctan a_n/b_n); thus we can plot a phase spectrum analogous to the amplitude spectrum.

When the series given by this equation is extrapolated beyond the range $(-\pi, \pi)$, the function $f(x)$ is repeated periodically in every interval between $(2n + 1)\pi$ and $(2n + 3)\pi$. It is possible to extend this analysis to a non-periodic function, provided that $\int_{-\infty}^{\infty} |f(x)| \cdot dx$ exists. The result is:

$$f(x) = \int_{-\infty}^{\infty} c(k) \cdot e^{ikz} \cdot dx$$

where

$$c(k) = \frac{1}{2\pi} \cdot \int_{-\infty}^{\infty} f(\xi) \cdot e^{-ik\xi} \cdot d\xi.$$

$c(k)$ is called the Fourier transform of $f(x)$. It is a complex variable, whose amplitude and phase can be plotted against frequency to give the amplitude and phase spectra of $f(x)$.

Thus, for example, if $f(x)$ represents the amplitude of a seismic source signal, we can use the amplitude spectrum to show how much of the energy of the source is available at different frequencies. A typical practical example is shown in figure 1/3. The power at any frequency is given by the square of the amplitude. Use of the Fourier transform allows us to pass between the time domain (that is, the actual signal) and the frequency domain (that is, the amplitude and phase spectra).

It is worth noting that there is an inverse relationship between the width of a signal and the width of its amplitude spectrum, so that a sharp spike has a flat spectrum and a very slowly varying signal has a sharply peaked spectrum. The function $e^{-\rho^2 x^2}$ demonstrates this relationship. As shown in figure 1/4, this is bell-shaped curve for which we can define a characteristic width by the x co-ordinate of the point where its value is $1/e$, which is given by $x = 1/\rho$. The Fourier transform of this function is given by:

$$c(k) = \frac{1}{2\pi} \cdot \int_{-\infty}^{\infty} e^{-\rho^2 x^2} \cdot e^{-ikx} \cdot dx,$$

$$= \frac{1}{2\pi} \cdot e^{-k^2/4\rho^2} \cdot \int_{-\infty}^{\infty} e^{-(\rho x + ik/2\rho)^2} \cdot dx,$$

$$= \frac{1}{2\pi\rho} \cdot e^{-k^2/4\rho^2} \cdot \int_{-\infty}^{\infty} e^{-z^2} \cdot dz, \text{ putting } z = \rho x + ik/2\rho.$$

It may be shown that the value of the integral is $\sqrt{\pi}$, giving

$$c(k) = \frac{1}{2\rho\sqrt{\pi}} \cdot e^{-k^2/4\rho^2}$$

Thus the Fourier transform is a bell-shaped curve of the same form as the original function, but the $1/e$ width is 2ρ. Thus a large value of ρ implies a sharply-peaked signal with a broad-peaked spectrum, and a small value of ρ implies the converse relationship.

Filters

It is easiest to think of filters as acting on the amplitude and phase spectra of a signal. The effects on both spectra must be specified for a complete statement of the action of a filter, although the effect on the amplitude spectrum is generally of most immediate concern. For example, in recording seismic data, we may employ an anti-alias filter

whose effect is to attenuate any signal or frequency over, say, 60Hz. The effect of such a filter can be represented by a function multiplying the amplitude spectrum as shown in figure 1/5.

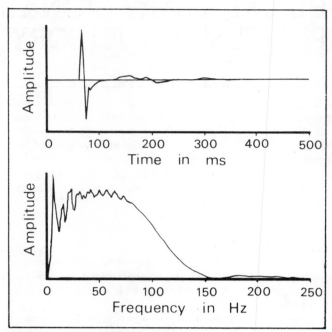

Figure 1/3: Typical seismic signal and its amplitude spectrum.

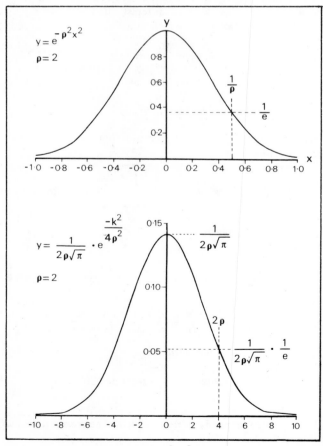

Figure 1/4: The function $y = e^{-\rho^2 \cdot x^2}$ and its Fourier transform.

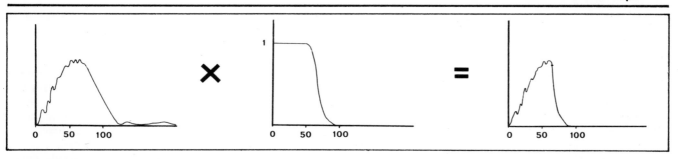

Figure 1/5: Filtering in the frequency domain: application of an anti-alias filter.

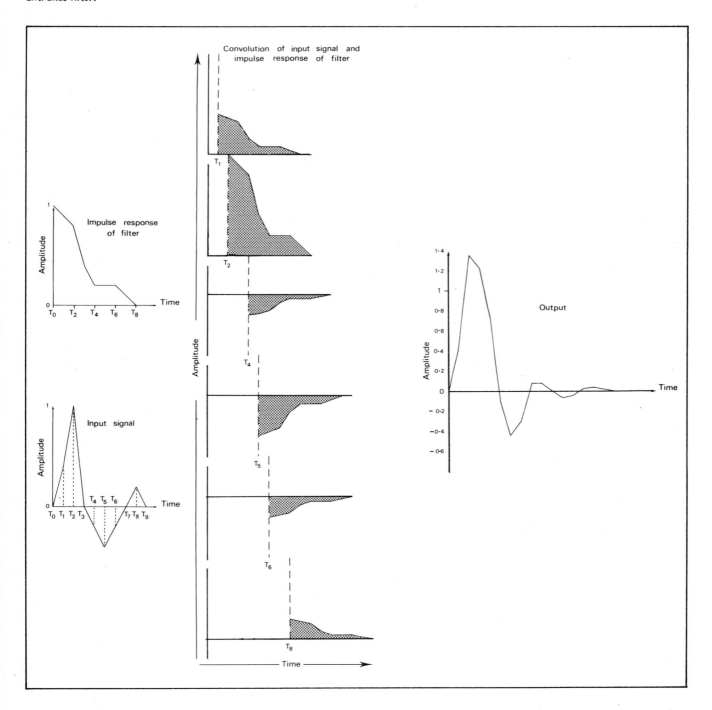

Figure 1/6: Generation of output signal for the impulse response of a filter convolved with an input signal.

187

Convolution

We would like to know the effect of a filter on the actual signal. One way to approach this problem is to use the Fourier transform to convert the filtered amplitude and phase spectra back into the time domain, but is is useful to have a way of thinking about the effect of a filter directly in the time domain. We can approach this by considering the impulse response of the filter, that is, the output when the input is a sharp spike at $t = 0$. If we know this function, then the effect of a filter can be simulated by breaking up the input signal into a series of such sharp spikes and adding the resulting impulse responses together (figure 1/6), with amplitudes proportional to that of the input signal element considered and time delays appropriate to that element.

Formally, this process can be expressed by means of the convolution integral. If the input function is $f(t)$ and the impulse response is $I(t)$, then their convolution is defined as:

$$\int_0^t f(\tau) \cdot I(t - \tau) \cdot d\tau,$$

and this is the required output function, often written as $f*I$.

We can pass at will between frequency and time domains, for it may be shown as follows that the Fourier transform of the convolution of an operator and an input function is the same within a scale factor as the product of the Fourier transforms of the operator and input function.

In cases of interest to us, $f(t)$ will be zero for times before some time $t = 0$ (the onset time of the signal), and we can without loss of generality make this to be the zero of time. Also, physically realiseable filters will not produce an output before any signal has been applied, so that $I(\tau)$ will be zero for negative τ. In this case the Fourier transform of the convolution can be written as:

$$F = \frac{1}{2\pi} \int_{-\infty}^{\infty} e^{-ikt} \cdot dt \cdot \int_0^{\infty} f(\tau) \cdot I(t - \tau) \cdot d\tau ,$$

since $I(t - \tau)$ will be zero for $\infty > \tau > t$,

thus $F = \frac{1}{2\pi} \cdot \int_{-\infty}^{\infty} f(\tau) \cdot d\tau \cdot e^{-ik\tau} \cdot \int_0^{\infty} I(t - \tau) \cdot e^{-ik(t-\tau)} \cdot dt,$

$= \frac{1}{2\pi} \cdot \int_0^{\infty} f(\tau) \cdot d\tau \cdot e^{-ik\tau} \cdot \int_{-\infty}^{\infty} I(u) \cdot e^{-iku} \cdot du,$

$= \frac{1}{2\pi} \cdot \int_{-\infty}^{\infty} f(\tau) \cdot d\tau \cdot e^{-ik\tau} \cdot \int_{-\infty}^{\infty} I(u) \cdot e^{-iku} \cdot du,$

since $f(\tau) = 0$ for negative τ, which except for a factor of 2π is the required product of the Fourier transforms.

Auto-correlation and cross-correlation functions

We may need to express the degree of similarity between two functions $f_1(t)$ and $f_2(t)$. One way of measuring this is to multiply the two functions together and integrate the resultant function; if the two functions are largely out of phase a negative number will result, and if there is no correlation between them the result will be near zero. To allow for phase lags between the signals we should repeat this multiplication and integration after shifting one signal with respect to the other by a particular time shift. Formally we can define a cross-correlation function $X(t)$ given by

$$X(t) = \int_{-\infty}^{\infty} f_1(\tau) \cdot f_2(t + \tau) \cdot d\tau.$$

If $f_1 = f_2 = f$, then we measure the similarity of f to itself at various time delays, and X becomes the auto-correlation function $A(t)$. Obviously the auto-correlation function will have its largest value at zero time-shift but if the signal has a dominant periodicity the auto-correlation function will show maxima at time shifts equal to this period and its multiples. In the case of the seismic signal, the auto-correlation function can therefore reveal dominant reverberations such as those between the seabed and the sea surface.

It can be shown that the Fourier transform of the auto-correlation function is the power spectrum of the original signal.

APPENDIX 2

American Association of Petroleum Geologists
Box 979
Tulsa
Oklahoma 74101
USA

Aquitaine Company of Canada Ltd
2000 Aquitaine Tower
540-5th Avenue SW
Calgary
Alberta
Canada

Atlantic Richfield Company
163/169 Brompton Road
London SW3
England

Banff Oil Ltd
(merged with Aquitaine Company of Canada Ltd in 1970).

Bolt Associates
205 Wilson Avenue
Norwalk
Connecticut 06854
USA

BP Petroleum Development Ltd
Farburn Industrial Estate
Dyce
Aberdeen AB2 0PB
Scotland

Compagnie Générale de Géophysique (CGG)
6 rue Galvani
9 1301 Massy
France

Decca Survey Ltd
Kingston Road
Leatherhead
Surrey KT22 7ND
England

Digitech Ltd
500, 441-5th Avenue SW
Calgary
Alberta
Canada T2P 2V1

Esso Australia Ltd
127 Kent Street
GPO Box 4047
Sydney
New South Wales 2001.
Australia.

Esso Exploration Inc
12727 Kimberley Lane
Box 146
Houston
Texas
USA.

Forest Oil Corporation
1500 Colorado National Building
950 - 17th Street
Denver
Colorado 80202
USA.

**Gas Council (Exploration) Ltd & Hydrocarbons Great
   Britain Ltd**
59 Bryanston Street
Marble Arch
London W1A 2AZ
England

Géomécanique
370 Avenue Napoleon Bonaparte
Rueil Malmaison
France

Geophysics
SEG Headquarters
P.O. Box 3098
Tulsa OK 74101
Oklahoma
USA

Geophysical Service International
Manton Lane
Bedford MK41 7PA
England

Geoquest International Inc
4605 Post Oak Place
Ste 130
Houston 77027
Texas
USA.

Geoterrex LTd
2060 Walkley Road
Ottawa
Ontario
Canada K1G 3P5

Grant Geophysical Corporation
7535 Flint Road SE
Calgary
Alberta
Canada T2H 1G3

Hamilton Brothers Oil Company (Great Britain) Ltd
Cleveland House
P.O. Box 17
19 St. James' Square
London SW1Y 4LP
England

Hunting Geology & Geophysics
Elstree Way
Boreham Wood
Herts
England

Institut Français du Pétrole (IFP)
1 & 4 Avenue de Bois-Préau
92 Rueil Malmaison
Hauts de Seine
France

Institute of Geological Sciences
Exhibition Road
South Kensington
London SW7 2DE
England

Mobil Oil Canada Ltd
330-5th Avenue SW
Calgary
Alberta
Canada T2P 0L4

Motorola Position Determining Systems
Stotford
Hitchen
Herts 8G5 4AY
England

Petty-Ray Geophysical
Coldharbour Lane House
106 Coldharbour Lane
Hayes
Middlesex
England

Philips Petroleum Company
Portland House
Stag Place
London SW1E 5DA
England

Phoenix Canada Oil Company Ltd
Ste 700
15 Toronto Street
Toronto
Ontario M5C 2E3
Canada

Prakla-Seismos GmbH
Postfach 4767 — Haarstrasse 5
3000 Hannover 1
West Germany

Sabre Petroleum Ltd
No. 1200, 444-5th Avenue SW
Calgary
Alberta
Canada T2P 2T8

S. &. A. Geophysical
S & A House
Azalea Drive
Swanley
Kent
England.

Schlumberger Well Services
5000 Gulf Freeway
P.O. Box 2175
Houston
Texas 77001
USA

Seiscom Delta Exploration Inc
Cumberland House
Fenian Street
Dublin 2
Ireland

Seismic Engineering Company
1133 Empire Central
Dallas
Texas 75247
USA

Seismograph Service Corporation (SSC)
P.O. Box 1590
Tulsa
Oklahoma 74102
USA

Seismograph Service (England) Ltd (SSL)
Holwood
Westerham Road
Keston
Kent BR2 6HD
England

Sercel
Laboratoires et Usine de Nantes — Carquefou
Cédex 25 X
44040 Nantes Cédex
France

Société Nationale des Pétroles d'Aquitaine
26 avenue des Lilas — 64 Pau
P.O. Box 65 64001
Pau
Cédex
France

Société pour le Développement de la Recherche Appliquée
  (SODERA)
9 bis rue Jean Mallard
83100 Toulon
France

Techmation
113-115 rue Lamarck
75018 Paris
France

Teknika Resource Development Ltd
412, 339-6th Avenue SW
Calgary
Alberta
Canada T2P 0R8

United Geophysical Corporation
2650 E. Foothill Blvd
Box "M"
Pasadena CA 91109
USA

Western Geophysical
Wesgeco House
288/290 Worton Road
Isleworth
Middlesex

LIST OF FIGURES

tape transport unit, field system 27, 32
terrain correction, gravity, gravity 110
tear fault 86
Texas Instruments DFS V seismic recording system 28, *2/25*
Thornburgh's wavefront method 98
thrust fault; *see* reverse (thrust) fault
thumper 13, 18
time average equation 2, 3, *1/4*
time break amplifier, field system 27, 31
time-depth conversion 161, 165, 166-170; *see also* depth
    conversion
time structure map, Kingfish 164, 165
time-variant filter 48
Toran 34
total depth (TD) 59, 64
total system controls, field system 27, 32
transcurrent faults 86
Transit satellites 34, *2/33*
transmission decay 9
Transverse Mercator projection 35
true amplitude: displays 57, *3/35, 8/6, 8/7;* processing 70;
    recovery 29, 119
two-way time 71, *4/4, 11/16;* map *75-76,* 79, *129, 5/6, 9/9,*
    9/10, 9/11, 9/12, 9/13, 10/14, 10/15, 11/3

unconformities 87-88, 115, 117, *6/10;* Kingfish oilfield 161;
    Rainbow Lake area 147, 148
Universal Transverse Mercator (UTM) 34, 35
uphole shooting 40

Vaporchoc 21, 25, *2/17*
variable area display 49, *3/23, 3/24*
variable density display 49, *3/24*
velocity: analysis 43-47, 70, 77, 112, 117, 165-171, *3/17, 3/18;*
    and depth map 75; Kingfish field 161, 163; map and grid
    76-78: P waves 112, 118; reef carbonates 116; in refraction
    work 40, 96, 97; of sound in sea water 38; in Wabanum unit
    148; *see also* average velocity, compensated sonic log,
    interval velocity, rms velocity, seismic velocity
vertical seismic profile (VSP) 64
Vibroseis 13, 16, 18, *2/11, 2/12*
Wabamun reflector 147, 148, 149, 152, *10/14*
'walk-away' survey 13
water gun 25, 91, *2/17*
wave equation migration 52-56, *3/29, 3/30, 3/31*
wavefront 7, 49, 51, 97, 98, *7/10*
wave-shaping kit 21
weight dropping 18
well velocity survey 59, 61-64, 66, 68, 168, *4/6*
Wentworth Classification 82, table *6/1*
Western Geophysical 21, 22
'wiggle trace' 49, 56, *3/23, 3/24*
Wind River Basin, Wyoming 122, *8/9*
wrench faults 86; *see also* strike-slip faults
Young's modulus 1, 2
zero minimum phase wavelets 66